The Making of the Himalayas through the Lens of Ladakh Geology, Structure, and Landscapes

This book provides a beautifully illustrated overview of the physical science of mountains, showcasing the evocative and highly informative terrains of Ladakh. Readers are taken across the mountains and valleys of Zanskar, the Indus Valley, and Nubra, relating a story that took millions of years to tell. This story includes the geology and tectonics of the Himalayas alongside the creation of landscapes, the people that live within them, and the scientific understanding required for minds to unravel the mysteries (mantras) of mountains.

The book is region- and theme-focused. Early chapters detail the science of continental collision tectonics, crustal deformation, and the geography of the Ladakhi mountains, together with a broad geotectonic overview of Ladakh and surrounding regions. Geology and stratigraphy of the Indian Plate units (High Himalaya Crystallines and Zanskar Sequence) are presented, step by step, together with deep crustal metamorphic processes, and the folding and thrusting of rocks, essential for the making of the Himalayas. The Kohistan-Ladakh Island Arc is presented as an opportunity to learn how new crust is created within oceans, producing batholithic and volcanic sequences, and how mountains are eroded to produce molasse sedimentary rocks. Geological accounts end crossing the Shyok Suture into the Karakoram Terrane. Focus then turns to how tectonics, ice, rivers, gravity, and wind shape and mould the spectacular mountain landscapes of Ladakh and the culture of resilient people who live amongst them.

This book is intended for anyone with a curious mind about mountains and how they are formed and shaped. Mountains evoke deep human connections to our dynamic Earth in humbling and inspiring ways. This book speaks to people within and way beyond academia and professional geology: to anyone who loves mountains.

The Making of the Himalayas through the Lens of Ladakh Geology, Structure, and Landscapes

Mantras of the Mountains

Michael G. Petterson
and Lanka Nanayakkara

CRC Press
Taylor & Francis Group
Boca Raton London New York

CRC Press is an imprint of the
Taylor & Francis Group, an **informa** business

Designed cover image: © Lanka Nanayakkara

First edition published 2027
by CRC Press
4 Park Square, Milton Park, Abingdon, Oxon, OX14 4RN

and by CRC Press
2385 NW Executive Center Drive, Suite 320, Boca Raton FL 33431

CRC Press is an imprint of informa UK Limited

British Library Cataloguing-in-Publication Data
A catalogue record for this book is available from the British Library

ISBN: 978-1-041-34717-0 (hbk)
ISBN: 978-1-041-34856-6 (pbk)
ISBN: 978-1-003-79486-8 (ebk)

DOI: 10.1201/9781003794868

Typeset in Times
by Deanta Global Publishing Services, Chennai, India

This book is dedicated to all sentient beings and our Dynamic Earth

For readers who have kindly purchased a printed book we ask you to please note that the narrative is drafted for a colour version of this manuscript, available as an electronic book. We ask you to bear this in mind and convert colour references to shades of black and white in your mind, when examining figures and reading text.

Contents

About the Authors

Michael G. Petterson, PhD, is from County Durham, North England. He has lived and worked in numerous countries through his geological adventures. Michael has been a Full Professor at two universities (Leicester, UK, and AUT, Auckland, NZ), is a Visiting Professor at the University of Ladakh, India, and Emeritus Professor at AUT, Auckland. He has been an Honorary Professor at the USP, Fiji Islands, and the University of Leicester, UK. He worked for many years with the British Geological Survey, culminating in the position of Director, and a member of the Executive Leadership team. Michael worked and lived extensively in the Pacific Islands, including holding the position of Director of Geoscience for the Secretariat of the Pacific Community (a UN-style development organisation). Michael has published extensively in academic and popular media. Much of his career has been spent applying geoscience to development in all its forms, particularly in the less privileged world. His love for all things Himalaya and Tibet began with PhD studies around Gilgit and Hunza, subsequently travelling and working in Afghanistan, Pakistan, India, Tibet, Nepal, Bangladesh, and Myanmar, observing the Himalayas in all their varied glories.

Lanka Nanayakkara grew up in Sri Lanka and was educated at Kelaniya and Colombo Universities. Lanka has worked as an information scientist with international libraries and has enjoyed travelling throughout Europe, Asia, and the Pacific, living in Fiji and New Zealand. The magical Ladakhi landscapes, their stories, and people deeply inspire her. An avid traveller herself, Lanka has an extensive photographic archive of Ladakh and has regularly posted highly researched pieces of writing and travel vignettes on social media. She has studied literature, sociology, linguistics, temple art history, and has a Master's degree in library and information science.

Forewords

Michael Petterson and Lanka Nanayakkara have meticulously studied Ladakh's landscapes and have successfully captured and articulated the true essence, scientifically and socially, of this remarkable region. During my tenure as Vice Chancellor at the University of Ladakh, my interactions with Professor Petterson revealed his passion and enriching knowledge about mountain systems and the Himalayan landscapes. His vision for this book, to reach a wider audience, is evident in the way the book presents an extensive multidimensional study, ranging from Ladakh's valleys, rock formations, and glaciers to the powerful influence of wind and other environmental forces. This book meaningfully and immensely contributes to the understanding of Ladakh's unique geography and environment and will undoubtedly be beneficial to a wide range of readers.

The goal of this book is to provide a comprehensive and forward-looking resource on mountain systems and the Himalayan landscapes. As the field continues to mature, such an integrated reference becomes essential for advancing evidence-based practice and responsible innovation. I believe the book will serve as a valuable companion for researchers and research development alike.

Prof. Dr S. K. Mehta, FRSC
Former Vice-Chancellor, University of Ladakh
Professor, Department of Chemistry, Panjab University, Chandigarh

As somone who was born and grew up in Ladakh, this book, for me, has been a prologue to a journey of endless discovery. The civilisational history interwoven with the rugged landscapes has always made me curious, leaving me restless to know more. Mantra of the Mountains has put me at ease by finally offering a holistic picture of what was, until now, an almost-myth about the continental collision that led to the three prominent mountain ranges in Ladakh, as well as valleys carved in between them by glaciers and rivers. This is a much-awaited work from someone who has put decades of dedication, with passion, to produce such an invaluable reference for future studies about the making of mountains in Ladakh.

Tashi Morup
Projects Director
Ladakh Arts & Media Organisation

Firstly, I extend my heartfelt congratulations on this beautifully crafted and soulful work. To me, this book is both an inspiring companion for enthusiasts and a practical handbook that illuminates the geological imagination with the clarity and resonance of mantras. It is a guide that not only nurtures creativity but also makes the art of imagination accessible to all who seek it.

The book is a masterfully crafted exploration of Ladakh, where the authors skilfully bridge the tectonic origins of the region with a profound sense of connectedness with its culture, traditions, and people. By gazing directly into the 'eye' of this amphitheatrical landscape, the writer challenges us to introspect – bridging the gap between the visible lithology and our own infinite cognitive imagination.

The book's ability to treat geological processes as a living language is incredible. Drawing from a wealth of experience, the authors translate complex geology into a simple, gripping narrative, making the earth's history accessible without losing its depth. They navigate the paleo-environments of the high plateau, showing how the relentless forces of glaciers, rivers, landforms and high-altitude winds shape sacred imprints upon our geological imagination.

From the depths of the Zanskar, Nubra, and Suru Valleys to the shores of Pangong and the ancient monasteries perched on the rugged scree across Ladakh, the authors meticulously document how geological activity – the folding, faulting, and the geological dynamic of Earth, – mirrors our own existence. It is a profound deliberation on how observing the naked terrain and the shifting cryosphere, with the assimilation of its culture, traditions, and people, can sometimes trigger significant 'waves of change' in our perspectives. This is a must-read for every enthusiast who wants to truly explore and understand the geology that surrounds us; it is a visceral reconnection with the deep time written into the very marrow of the Ladakh landscapes.

At last, I wish you both every success with this book, where the hard work, devoted fieldwork, hours of deliberation and long nights of reflection converge into a beautiful culmination. It is not merely a work, but a pilgrimage etched with imprints like mantras on stone, alive with the interdependence of people, places, and things encountered along this sojourn. Within its pages, bound by threads of interconnectedness, memory and imagination entwine, creating a tapestry at once personal and universal, intimate and with infinite possibilities.

Lobsang Palden
Geologist, Government Department of Geology
UT of Ladakh, India

Recent advances in geoscience have resulted in the progressive deepening and diversification of the discipline into narrowly specialised research areas. This trend has been particularly intensified by the rapid development and refinement of modern analytical instruments used to investigate rocks, minerals, nanoparticles, elements, and isotopes. Although these disciplines advance our understanding of fundamental physical and chemical processes in geoscience, their increasing specialisation has led to a growing isolation from other geoscientific fields. Consequently, their connection to the natural beauty of geology, which often motivates our engagement with the field, may not be immediately apparent. Thus, any attempt to bridge these specialised disciplines with the geological processes that govern the formation of geological structures and landscape morphology is particularly valuable. It is crucial to convey a perspective that appeals not only to professionals in related fields but also to students who are contemplating their future paths. These expectations seem to have been fulfilled by the book *Mantras of the Mountains* by Michael Petterson and Lanka Nanayakkara, which explores the geology and related building processes of the Himalayas and Ladakh.

The book begins with an overview of the origin and formation of the Himalayas, supported by well-illustrated maps that depict the tectonic units produced by the collision between the Indian subcontinent and Eurasia. It also addresses how climatic changes have influenced the shaping of the Himalayas and contributed to the emergence of their characteristic natural scenery. Other chapters provide a more detailed account of the geology of Ladakh, with emphasis on lithology, landscape development, and glacial and hydrological processes. The text is illustrated with cartoons, geotectonic models, and landscape imagery that not only explain the geodynamic evolution of the Himalayas, with a particular focus on Ladakh, but also provide a valuable background for geological excursions, and a chapter on links between mountains, tectonics, and culture, which elucidates how, as humans, we are influenced by geological forces. It includes details on the lithological variability of rocks and their structures, further illustrated with hand-drawn schematics highlighting these variations and structural features. The picturesque images of high mountains, valleys, rivers, streams, and glaciers captivate both the general public and those with a keen interest in nature.

Professor Shah Wali Faryad
Charles University, Prague
Czech Republic
World-renowned researcher into metamorphic rocks and mountains,
with great experience in the Afghan Hindu Kush

Preface

A narrated pictorial journey across Zanskar, the Indus Valley, and Nubra: from the High Himalaya to the Karakoram

The mystical former kingdom of Ladakh, NW India, bordering the western periphery of Tibet, steeped in Tibetan Buddhist culture, is a wonderful showcase in which to understand the genesis and evolution of mountains. This book takes the reader, step by step, through a world of tectonics, mountain-building, and Ladakhi spectacular scenery. Numerous photographs and illustrations communicate the subtleties and secrets of the rocks and landscapes of Ladakh. The accompanying narrative uses illustrative material to weave a mountain tapestry: moving from the plate tectonics of the Himalayas to the geological materials that make up the mountains, the landscapes that result from ice, water, and wind, and a flavour of the people who evolved with the mountains.

Ladakh is a high-mountain desert, with even the lowest-ground being above 3000 m, presenting thousands of square kilometres of rock outcrops, valley floors, and mountain cliffs and ridges, alongside a myriad of Himalayan peaks. This book unravels the three- and four-dimensional space-time geological story of an Indian continent, a sandwiched island arc, and the tectonics of Central Asia. Starting from first principles, with scientific terms explained in common language, the reader is taken on a journey through the mechanics of continental collision and the resultant architecture of the Himalayas, the Trans-Himalayas, and Tibet. Each geographical zone of Ladakh,from the High Himalaya southern border of Zanskar and Himachal Pradesh, moving northwards through the Zanskar Mountains, and onwards across the River Indus to Nubra, Pangong, and the Karakoram mountains, – is described and explained in detail, through pictures and words. Key geological units such as: the High Himalaya Crystallines, pushed up from the deep crust, many kilometres beneath the surface; the Zanskar Sedimentary Sequence, folded and contorted into amazing, weird, and wonderful fold shapes; and an entire island arc terrane composed of volcanic and plutonic igneous rocks, including a huge batholith or aggregation of granite, that sustains the very continental crust on which we all live, are beautifully exposed and presented.

Ladakhi landscapes vary from broad valley floors to deep gorges, wide mountain ranges, incised 'U' and 'V' shaped mountain valleys, and amphitheatre bowls carved high in the Himalayas. Ice accumulates as glaciers and rapidly carves deep and wide valleys, depositing vast quantities of eroded boulders, gravels, sands and muds. Rivers are strengthened in power through the rapid rise of mountains, allowing them to cut scarily impressive, sheer walls through vertical incisions. Alluvial fans are ubiquitous in this mountain desert environment, spreading a jumble of deposits comprised of huge boulders, embedded in sand and mud, extending from the mountain front to the plains, on top of which many Ladakhi villages are built. Wind sand-blasts ridges, mountain slopes, alluvial deposits, and rocks, alongside the accumulation of great thickness of sand in natural bowls and amphitheatres, the creation of ripples and dunes, and the production of an impressive array of natural sculptures. Tectonics produces sharp linear escarpments and broad valley floors. The last chapter examines how mountains shape people, their resilience, and their approach to difficult lives in remote environments.

This is a scholarly work, grounded within literature and science foundations. Key papers are cited, with a focus on maintaining a continuous, interesting narrative, closely linked to presented photographs, figures, and diagrams. The intention is to make the book accessible to as wide an audience as possible, whilst deeply informing and educating.

The book takes the reader through a Himalayan Shangri-La: Ladakh, the land of the mountain passes, hundreds of millions of years of time, and the net impact of tectonic plates, volcanoes, disappearing oceans, and grinding faults, one step at a time, unlocking the mantras of the mountains. Mantras are deep and mystical, with a power to take us deeper and deeper into a world of greater understanding and knowledge. Likewise, learning to read the runes of the mountains and the language they speak to each and every one of us profoundly deepens our appreciation of one of the most magnificent manifestations of our dynamic planet. Without this dynamism, humanity and all of the living beings on our planet would not have appeared. The interdependencies between planet, process, time, and life are here for all to see. But unravelling the mountain mantras is the only way to see what we may never have ever seen before in our lives. Perhaps this book can help.

Michael G. Petterson and Lanka Nanayakkara

Acknowledgements

As authors, we relied heavily on travel and research. It would not have been possible without the support and help of many Ladakhis. Our first and greatest thanks go to the indispensable resource Palden, for his invaluable local knowledge in all things geological and logistical, his time, and wonderful companionship. To Yangjor, for driving, navigating, and guiding us throughout Zanskar, up impossible mountains to remote valleys, and for bringing us in contact with fascinating locals full of enchanting tales. His stories about yaks, dzos, bears, and Ladakhi cultural nuances and traditions helped us get a sense of this vast and ancient region.

To Rigzen, who drove and guided us so expertly to many corners of Nubra Valley, Indus Valley, and Leh, and for his great knowledge of Ladakhi wildlife, especially the snow leopard.

To Rahul in Leh for his friendship, deep conversations, and hospitality, so essential when one is in a foreign country. To Jack Vong, with whom we first made acquaintance with in Ladakh in the 1980s, with whom we share a deep passion for Ladakh, and for putting us in contact with T A Angmo and her family, who very readily shared their huge experiences of Ladakh and warm meals in cold December days.

While researching for this book, we stayed in numerous homestays and accommodations across Zanskar, Leh, Nubra Valley, and Changthang. We can't possibly name everyone and everywhere, but would like to thank the following people for shelter, respite, and laughter. We particularly would like to express our appreciation to the owners of Guru View Homestay in Pibiting. To Sherab for feeding us with delicious fresh vegetables from her garden and making sure we had copious amounts of butter tea; to Tenzin for being such a wonderful host and making us part of the family, and to Abi-ley (grandad) for entertaining us with interesting tales from old Ladakh. To the Grand Dragon and staff who greatly enhanced our experience in Leh, and offered professional service and luxury in abundance. It is thus fitting that acknowledgements go to the Abdu family and all the wonderfully helpful Grand Dragon staff. The hotel was our very comfortable base in Leh. To Tsering Palmo of Dorjey Namgail Homestay in Lingshed and the oral histories she shared with us.

We thank owners and staff of Eco-Lodge, Karsha, La Himalaya, Rangdum Guest House, Dolma Camp, Phuktal, Mohammod of Suru Valley Resort, the Dunes Hotel Hunder, Dragon Hotel Lamayuru, Gombo Camp, and Pangong Guest House.

To Tashi Morup and the staff of LAMO (Ladakh Arts and Media Organisation) in Leh for inviting us to give presentations, share experiences, and meet artists and scientists deeply interested in Ladakh.

We are grateful to the staff of the Science Faculty of the University of Ladakh, main campus, Taru, who gave us their time, especially VC Professor S.K. Mehta, Professor Subrat Sharma, and Dr Ibrahim. Thanks also to Graham Hinchliffe from AUT University, Auckland, New Zealand for his constant technical support, patience, and advice.

We are also thankful to many locals who invited us into their homes, and nuns and monks who offered endless cups of namkeen chai and shared local history within their Monasteries and Temples.

Michael wishes to acknowledge and thank Auckland University of Technology for providing a generous research grant and study leave while researching for, and writing this book.

We would like to express our profound gratitude to the Tibetan Buddhist Centre, Dorje Chang Institute (DCI), Auckland, for their deep kindness and spiritual support. Thank you, Geshe La Lharampa Thupten Wangchen and Venerables Gyalten Wangmo and Caroline.

To Rinchen at DCI for his invaluable friendship, great sense of humour, wisdom, encouragement, historical knowledge of Tibet and Tibetan Buddhism, and never-ending well-wishes.

We thank the detailed reviews from a number of anonymous reviewers, who have significantly improved the shape, form, structure, and style of this book.

Glossary

Ablation Zone. Area of a valley glacier where ice melting exceeds ice accumulation. Areas of crust that subside due to the overlying mass of growing mountains, accommodating sediment derived from the erosion of proximal mountains.

Abrasion. Erosion process that physically erodes rock surfaces and particles through the action of clast-on-clast interactions within currents and glaciers.

Amphibolite Grade Metamorphic Rock. High-grade metamorphic rock that still retains an element of hydrous minerals.

Andean Margin. Continental plate being subducted by an oceanic plate and experiencing earthquakes magmatism and volcanism.

Antiform. An upwards-closing fold structure.

Aquifer. Rock strata that contain groundwater resources.

Arrête. A sharp mountain peak produced through glacial and freeze-thaw action.

Attrition. Rock particles wear themselves down, with the rounding of sharp edges, as they are transported downstream by wind or water.

Axial Plane. A 3-D planar structure that bisects folds.

Bedding. 3-D penetrative structures formed during sedimentation and later 'hardened' through lithification. These primary structures are originally horizontal but may later dip through tectonism.

Bioturbation. (pertains to) organisms that mix sediments within a soft substrate, e.g. in shallow marine environments.

Boudinage. Rock deformation characterised by layers being broken up into discrete pods or boudins.

Boulder Clay. (pertains to) glacial deposits, which are very poorly sorted and often contain boulders alongside clay.

Braided Rivers. River channels that resemble braided hair or jewellery, with numerous channels separated by 'islands' of sand and gravel.

Breccia. A poorly sorted rock, similar to a conglomerate, but with angular, larger clasts (rock fragments). Larger clasts are embedded within a finer-grained matrix/groundmass.

Brittle Deformation. Deformation typified by breaking rocks, creating faults and fractures, normally at lower temperatures.

Chert. Silica-rich sedimentary rock, formed in the deep ocean.

Chevron Folds. Zig-Zag folds that resemble chevron shapes.

Chilled margins. Rapidly chilled edges to igneous intrusions with smaller crystal sizes.

Cleavage. Penetrative planar structure within a metamorphic rock along which a rock may be parted. Formed through preferential alignment of metamorphic minerals.

Clast. A discrete rock particle or fragment.

Clast Imbrication. Alignment of clast long axes in response to current flow.

Coarse-Tail Graded Bedding. Grading that only affects the larger clasts within a sedimentary or volcaniclastic rock

Conglomerate. Sedimentary rock, which is very poorly sorted, containing large rounded clasts set within a finer grained matrix.

Corrie. A bowl-shaped topographic structure within mountains formed by the accumulation of ice. Corries host the accumulation zones of glaciers.

Cross Bedding. Localised bedding which pertains to an angle to the dominant bedding within a sedimentary rock. Indicates the direction of currents which deposited sediments.

Dendritic River Network. River drainage pattern that resembles a tree in plan view.

Desert Pavements. Desert ground surfaces depleted in finer particles.

Desert Varnish. Exposed rocks develop yellow-brown-ochre-red 'skins' of oxidised and hydrated minerals within desert environments.

Detachment Surface. A low-angle fault that separates upper geological blocks above the detachment surface from geology below. Detachments can facilitate movement over many tens of kilometres.

Dip. Geological measurement of inclined strata that indicates the degree of strata tilt relative to the horizontal and the compass bearing of the dip.

Dip Slope. Escarpment slope with one surface dipping close to the dip angle of underlying geological units

Disconformity. A geological structure that separates geological units with a mismatch in structural integrity.

Disharmonic folds. Folds with a complex geometry and a change in shape throughout the structure.

Dissolution. Process which dissolves minerals and rocks, and through which rocks are eroded and transported by rivers and oceans.

Ductile Deformation. Deformation characterised by plastic deformation of rocks at higher temperatures.

Dyke. A sheet-shaped intrusive igneous structure with a vertical or high dip angle disposition.

Effusive Volcanic Eruption. Relatively quiet volcanic eruption.

Escarpment. Topographic feature with a steep (scarp) slope and gentler (dip slope) that meet at a central crest.

Eskers. Sinuous sand and gravel ridges of glacial-fluvial origin.

Eutaxitic Texture. A pyroclastic rock with aligned fiamme.

Explosive Volcanic Eruption. Relatively violent volcanic eruption.

Extensional tectonics. Tectonics that pull crust apart and form rift valleys.

Facies. pertains to Environments and associated conditions that produced the rock under study e.g. sedimentary rocks may have a shallow marine facies.

Fault. Planar discontinuity within geological strata where movement has occurred either side of the discontinuity (fault plane).

Fiamme. Flattened pumice clasts within a pyroclastic rock.

Fissile. A rock capable of being easily split along planes of weakness is a fissile rock.

Flower Structure. Geological structure with a lower 'stem' and upper 'branches' that resemble trees and flowers. Formed within compressive thrust fault stress regimes.

Fold Axis. Linear structure that marks the trace of fold closure (antiforms and synforms) across terrains.

Fold Limbs. Areas of folds that connect synforms to antiforms and vice versa. All folds have two limbs.

Foliation. 3-D planar penetrative metamorphic fabric within a metamorphic rock.

Foreland basin. Zone of crustal depression in front of active mountains that receives and accumulates sediment in a basin morphology.

Freeze-thaw. (pertains to) the weathering action of freezing and thawing cycles that break up rock masses.

Geothermal gradient. Rate of temperature increase with depth in the Earth.

Glacial plucking. (pertains to) erosive action of glaciers that freeze ice onto rock surfaces and subsequently detach them through glacial movement.

Glacio-fluvial. (pertains to) processes that combine ice and water activity within glacial environments.

Gneiss. Coarse-grained foliated crystalline high-grade metamorphic rock.

Graded Bedding. (pertains to) rocks that display a systematic change in particle size with stratigraphic height. Normal grading fines upwards, whilst reverse grading coarsens upwards.

Granulite. High grade metamorphic rock comprised of anhydrous minerals.

Greenschist. Low-moderate grade metamorphic rock often containing green indicator minerals such as epidote.

Groundmass (or matrix). Finer-grained, mostly majority, 'background' of particles or crystals within a rock.

Hanging valley. A valley formed by a small tributary glacier that never reached the main glacier and is left 'hanging' above the main glacier, once ice has melted.

Heterolithic. A sedimentary rock that contains particles of different lithologies.
Hydraulic action. (pertains to) the sheer force of moving currents (atmospheric, river, ocean) that can erode and transport rock particles.
Igneous Rock. Rock which formed from the cooling of liquid magma.
Ignimbrite. A welded pyroclastic rock that results from hot pyroclastic flow eruptions. Comprised of fiamme, pumice, crystals, glass shards, lithic clasts, and ash.
Indentation tectonics. Style of tectonics where one continental plate indents another. E.g. India into Eurasia.
Interlocking Spurs. Valleys with a winding profile characterised by ridges interdigitating with one another. These valleys are typical of river activity in their upper courses.
Interstadial. A warmer spell within an ice age.
Intrusive Rock. Igneous rock which froze at depth and 'intruded' into crust.
Island Arc. New buoyant continental crust produced through the subduction of one ocean plate under another. A series of islands forms which have an arc geometry from a bird's-eye view: hence, the term 'island arc'. Examples include the Lesser Antilles, the Kermadec-Tonga arc, Vanuatu, Solomon Islands, South Sandwich Islands, Kurile Islands.
Isoclinal Fold. A fold with parallel fold limbs.
Lava. Liquid magma flowing like rivers from a volcano.
Lateral Moraine. Glacial sediments formed at the sides of moving glaciers.
Leucocratic. Igneous or metamorphic rock with a preponderance of light-coloured minerals.
Leucogranite. A granite with less than 5% dark minerals. Mostly formed by crustal melting. Orogenic belts commonly produce leucogranites, as in the Himalayas.
Lithology. (pertains to) type of rock.
Lithification. Process of sediment turned into hard, solid rock.
Mantle Plume. A buoyant balloon-shaped large volume of material hotter than the surrounding mantle, which rises through the mantle, breaks through the crust, and can erupt large volumes of basalt lava.
Mass wasting. (pertains to) downward movement of scree and soils through gravity on steep slopes.
Medial Moraines. Glacial sediments that accumulate within the centre of moving glaciers.
Melanocratic. Igneous or metamorphic rock with a preponderance of dark minerals.
Melange. A tectonic breccia is often located within suture zones.
Mesocratic. Igneous and metamorphic rocks with equal percentages of dark and light minerals.
Metamorphic Grade. (pertains to) the extent of metamorphism a rock has undergone and the maximum temperatures-pressures it experienced.
Metamorphic rocks. Rocks changed through heat and pressure in the solid state.
Micrite. A fine-grained crystalline limestone.
Migmatite. High-grade metamorphic rock that has undergone partial melting.
Misfit River. A river that now occupies a wide glacial valley, post-ice melt.
Molasse. Tectonic breccia/conglomerate/sandstone etc., formed within eroding rising mountains
Moraine. (pertains to) deposits formed from glaciers. Generic term.
Nappe. Flat-lying isoclinal fold. Formed as part of orogenesis. Nappes thicken and translate crust.
Normal Fault. A fault where one fault block is downthrown relative to another, across a fault plane. Normal faults result from extensional shear.
Obducted. (pertains to) tectonics where one geological unit is obducted on top of another, particularly as plates collide, and one plate obducts onto the other.
Ophiolite. Obducted ocean floor onto the continent.
Orogenesis. Tectonics of mountain building.
Orogenic Gold. Term for a style of economic mineralisation linked to orogenesis.
Pahoehoe Lava. A structure that describes rope-like structures which form in low-viscosity lavas, normally basalts.

Passive margin. A tectonically quiet junction between ocean and continental crust, usually one side of a passively diverging, expanding ocean.

Phenocryst. Minerals that form the larger crystals within an igneous porphyritic rock.

Pillow Lava. A lava that erupts or travels subaqueously and forms hemispherical 'pillow' shaped structures.

Pinch and swell structures. Deformed beds that thicken and thin.

Porphyritic texture. An igneous rock with large minerals set within a finer-grained matrix.

Porphyroblastic. Metamorphic rock texture with large minerals set within a finer-grained matrix.

Porphyroblast. Minerals that form the larger crystals within a metamorphic porphyroblastic rock.

Preferred Orientation. (pertains to) a metamorphic rock texture comprising aligned minerals.

Protolith. Geological material (rock) prior to being metamorphosed.

Pure Shear. Shear formed through direct push or pull between two forces: e.g., pushing a piece of plasticine in the same plane through which it compresses and thickens, or is pulled apart.

Pyroclastic fall. (pertains to) volcanic clasts that form from eruptions that launch them into the atmosphere and are later deposited from the atmosphere.

Pyroclastic flows. Volcanic flows comprising particles and gas at high temperatures that flow at high velocities and have very low viscosities.

Pyroclastic Rock. An igneous rock made of igneous clasts (crystals, ash and glass particles) derived from an explosive pyroclastic eruption. Pyroclast literally means a rock clast formed by heat.

Radial River Network. River drainage pattern that radiates from a central point. Like the spokes of a wheel.

Reaction Rim. A mineral texture around the rim of a mineral formed of new minerals produced through reactions between the mineral and the surrounding magma or metamorphic environment.

Rheology. (pertains to) Rock mechanical properties.

River catchment. Area of landscape that supplies all rivers with water/precipitation within the catchments, separated from other catchments by watersheds.

River courses or stages. (pertains to) the changing character of rivers and their environments as they travel from the mountains to the sea.

River long profile. (pertains to) the shape of a river course cross-section from source to mouth. Ideal river profiles approximate a concave-upwards parabolic curve.

River nick points. Sudden breaks in river long profiles caused by tectonic uplift and consequent readjustment of river long profile.

Rift Valley. A valley bounded by extensional normal faults.

River Rejuvenation. Rivers that experience tectonic uplift, which in turn increases the erosive power of rivers.

Saltation. Parabolic 'hopping' of rock particles by moving currents of air or water.

Scarp slope. Steepest escarpment slope.

Schist and schistosity. A metamorphic rock with a schistose fabric. A schistose fabric is a strong 3-D penetrative metamorphic planar fabric with mica as a dominant fabric-forming mineral.

Scree and Talus. Terms used for boulder-cobble-pebble sized clasts eroded from steep mountain slopes and accumulating on mountain slopes and bases of mountains.

Sedimentary Structures. Structures within sedimentary rocks that provide clues about their original formational sedimentary environment: e.g., ripple structures.

Shear. Rock mass undergoing deformation through applied tectonic force.

Shear Vector. Direction of shear.

Shear Zone. Zone of highly sheared rock, exhibiting strong plastic deformation.

Sill. A sheet-shaped intrusive igneous structure with a horizontal disposition.

Simple Shear. Shear produced through two forces pushing or pulling in different planes: e.g., pushing the top and bottom of a deck of playing cards in opposite directions.

Slate. Fine-grained low-grade metamorphic rock with a strong cleavage .

Slickensides. Surface structures produced on fault planes, caused by fault movement friction, that indicates fault motion vector.
Solidus. A rock begins to melt when the solidus temperature is exceeded, from c. 650 to 900°C.
Sorting. (pertains to) the variation in sedimentary rock clast size. A well-sorted rock is like a bag of sugar; a poorly sorted rock is like a pack of muesli.
Sparite. A coarse-grained crystalline limestone.
Stadial. A colder period within an ice age.
Strike. Geological measurement of line of zero dip on a rock outcrop layer.
Strike Slip Fault. Fault with horizontal-only motion.
Suspension. Particles suspended and transported in moving currents of air and water.
Suture. Fault that bounds two formerly separated geological terranes or plates.
Synform. A down-closing fold structure.
Terminal moraine. Glacial sediment that accumulates at the front edge of a glacier.
Thrust fault. Low-angle reverse fault caused by compression.
Till. Generic term for widespread deposits of glacial origin.
Tillite. Lithified till.
Traction. (pertains to) rock and mineral particles 'dragged' down-flow across a ground surface by moving currents of wind and water.
Trellis River Drainage. River drainage pattern that is rectilinear, similar to a trellis fence in shape/plan view.
Truncated Spurs. (pertains to) glacial valleys produced by glaciers eroding away all side-ridges, producing straightened valley profiles.
Tuff. Welded and lithified pyroclastic volcanic rock.
Turbidite. A rock formed from subaqueous sediment-charged dense water currents that are gravity driven and ground-hugging. Deposits are often normally graded and vary between conglomerate and mudstone.
'U' shaped valley. A glacial valley with steep valley sides and a wide valley floor.
Ultrabasic Rocks. Igneous rocks formed only of dark Fe-Mg-rich minerals have a low SiO_2 content and mostly originate from the mantle.
Unconformity. A geological structure that separates older units from younger units with a significant time gap between the two units. Classic unconformities separate the upper horizontal strata from the lower folded strata.
'V' shaped valley. A valley with steep sides and a narrow valley floor formed from the down-cutting of rivers.
Vesicles. Holes within lavas produced by lava degassing on cooling.
Volcaniclastic Rock. Volcanic rock formed from clasts.
Watershed. Line in a landscape that separates catchments and channels water down one side or the other (of the watershed).
Way-Up-Criteria. Structures that indicate the original 'right-way up' of a rock when formed (tectonics can turn rocks upside down).
Weathering. (pertains to) Earth surface processes that degenerate exposed rocks previously buried through physical, chemical, and biological processes.
Welded tuff. A tuff that has formed through the welding of volcanic pyroclastic clasts.

Tectonics and Geographical Setting of the Himalaya-Tibet Region

1

The Himalaya conjures up numerous images in people's minds. Unfailingly, the word 'Himalaya' evokes pictures of majesty, towering peaks, ice, and grandeur. To many, 'Himalaya' creates romantic, mystical, even magical mental responses. Terms such as 'Shangri-La' and 'Shambhala' have been used to articulate impressions roped in mystique, marvel, and the otherworldly, in response to outsiders unfamiliar with the region, painting atmospheres that dress the region in romance (Hilton, 1933). Humans regularly ask, 'can heaven on Earth exist within these mountains'? Surely this is the closest we can get to the spirits beyond this Earth? Stories of Tibet and other parts of the Himalaya during times of low information fed open and imaginative minds. Tintin comics animated Tibetan Shangri-La-like sentiments and meetings of spiritual beings (Hergé, 1960). Tibet was a closed and insular world for centuries, discouraging contact with the world beyond. This 'secrecy' added to the mysterious, unknown western picture-maps of the region. As time passed, more information became freely available to the world beyond Tibet and its regions. The *GreatGame* between the Russian and British Empires led to intrepid spies wandering through the regions of Ladakh, Kohistan, Afghanistan, Karakoram, and beyond. The British Younghusband expedition to Tibet in 1903/4 'persuaded' Tibet to (briefly) open up in limited terms with British India (Younghusband, 1910). Geographical and geological explorers from the late 19th and early to mid-20th century wrote reports and drafted maps of numerous Himalayan regions. Post-colonial India-Pakistan changed the Indian subcontinental governance arrangements, fell into warfare over border regions, and undertook further investigations of the Himalayan regions.

The Chinese invasion of Tibet closed this region to outsiders for decades. The breakup of the Soviet Union led to the creation of new 'stans' in the NW Himalaya, and independent Mongolia to the north of Tibet. Geological research ramped up from the 1960s/1970s and opened new data and vistas of the Greater Trans-Himalayan region, and, later, Tibet. The advent of the satellite age, particularly with the development of highly advanced sensors and high-resolution imagery, took away much of the mystique, replacing it with 'hard facts'. Travellers to Ladakh, Nepal, northern Pakistan, and later, Tibet, unravelled narratives from numerous angles. As the factual foundation for the Himalaya grew, the search for a mystical 'Shangri-La' and 'Shambhala' became more elusive. Realities of often-harsh human existences within extreme mountain environments became better known, although magic indeed remains.

Figure 1.1 presents the full extent of the Greater Himalaya-Tibet region. Geographers debate endlessly over definitions of the word 'Himalaya' including where it officially ends and begins. With justification, academics attempt to tightly define what nature itself defies. In reality, mountain chains run into each other and are rather seamless. Humans like to define, name, and map to help human minds understand natural complexity. Figure 1.1 presents an array of names such as 'Himalaya', 'Tibet', 'Karakoram', and 'Tien Shan'. Figure 1.1 demonstrates that the impact of the India-Eurasia tectonic collision goes way beyond the Himalayan arc that marks the surface boundary of active tectonics. The collision has had a

DOI: 10.1201/9781003794868-1

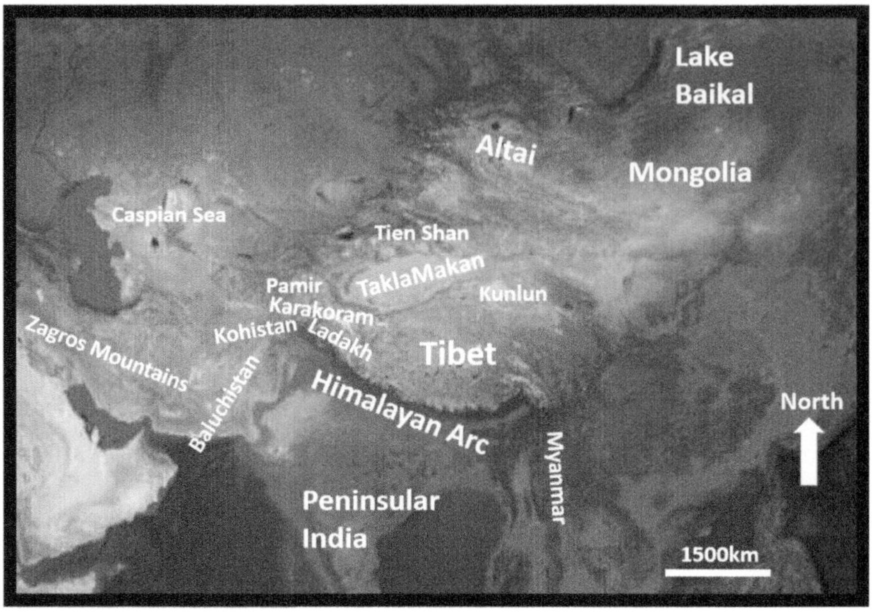

FIGURE 1.1 Geographical setting of the Himalaya-Tibet region. *Source*: Author's diagram. Base map acknowledgements: Google, N.D. Map of the Himalayan Region. Retrieved 20 February 2026, from https://goo.gl/maps/@29.321388,84.4468416,3968807m/data

pre-eminent or major role in the creation of the modern architecture of a huge mountain and basin system, present throughout central-eastern and south-eastern Asia. 'Himalaya' can, arguably, be restricted to the southern boundary of Tibet, pinned between the >8,000 m high Nanga Parbat Mountain in northern Pakistan, running through NW India and Nepal, and terminating in northern Myanmar. North of the 'Himalaya' is the Tibetan Plateau, with an average elevation of 5,000 m above sea level (asl). The North Western (NW) mountains in Ladakh, Kohistan, Karakoram, and Pamir are situated north of the more precisely defined 'Himalaya'. The 'grain' of the Greater Himalaya changes from NW-SE and east-west to north-south in Baluchistan and Myanmar, marking the western and eastern edges of the Indian Plate, respectively. The term 'Trans-Himalaya' (Hedin, 1909) is a term sometimes used for the NW mountain chains, north of the 'main-Himalaya'. Further north and east of the Trans-Himalaya are mountain chains such as Kunlun, Tien Shan, and Altai, relative topographic depressions that form the Taklamakan desert, and the deepest freshwater lake in the world (Baikal in Russia). This trans-continental physical architecture has been created in part or whole by the complexities of tectonics that resulted from India colliding with Asia. Whatever the whys and wherefores of strict definitions of mountain belts and geographical areas, the simple truth is clear from Figure 1.1: there is a diverse and grandiose continental fabric bounded by Indian Plate geometry to the south, and responses to the Himalayan collision throughout the Asian hinterland (Molnar and Tapponnier, 1975; Tapponnier et al., 1982). The highly fertile Indo-Gangetic Plain, sandwiched between the Himalaya to the north and Peninsular India to the south, owes its existence to the mountains. These plains are the bread and rice basket to more than 1 billion people, formed through mountain erosion and transport of sands, gravels, boulders, silt, and mud, accumulated within deep sedimentary basins, providing the substrate for the fecund Indo-Pakistan plains. To go deeper in our understanding of the creation of the Himalaya and Tibetan regions, we must (counter-intuitively) break up supercontinents, destroy large oceans, and crash plates together: a time journey of around 100 million years (Figure 1.2).

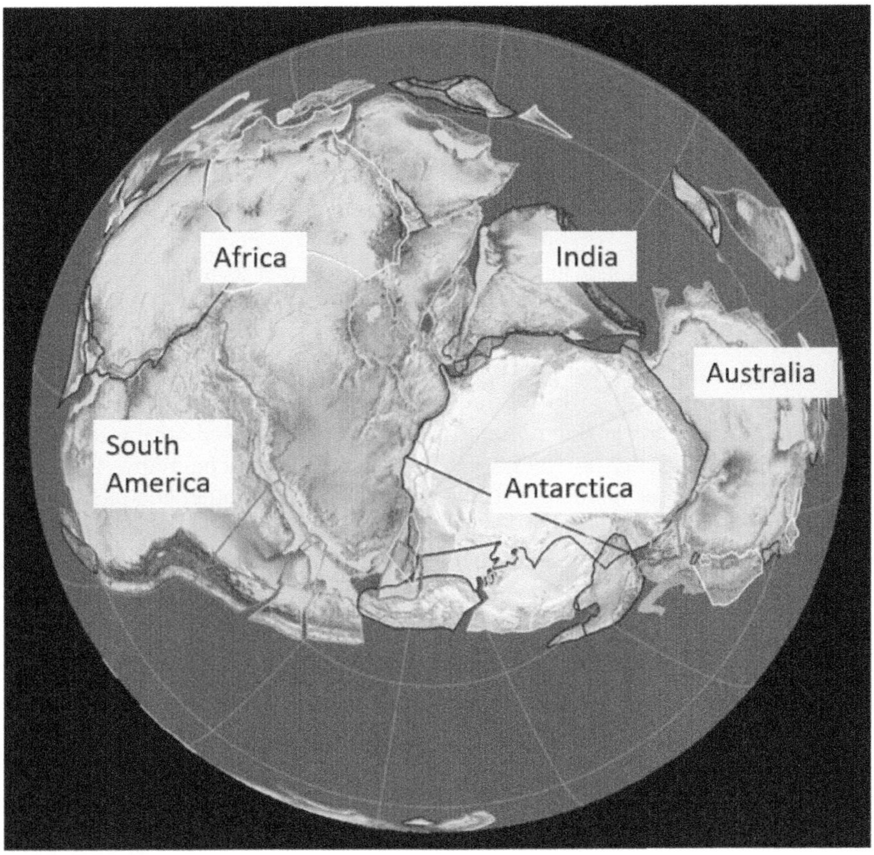

FIGURE 1.2 Map of Gondwana Supercontinent from 420 million years ago (Ma). *Source*: Acknowledgement. https://commons.wikimedia.org/wiki/File:Gondwana_420_Ma.png. Clamosa, 2018. Gondwana at 420 Ma. View centred on the South Pole. Shared through Creative Commons licence CC by 4 international attribution. https://creativecommons.org/licenses/by/4.0/. Annotations were added to the original.

BREAK-UP OF A SUPERCONTINENT AND NORTHWARDS DRIFT OF INDIA

The time journey begins around 600 million years ago (Ma) when a new southern supercontinent called Gondwana, Figure 1.2, (Suess, 1883), was created. The modern continents of Africa, South America, Australia, Antarctica, and India were all part of this supercontinent. The story of plate tectonics is one of continents dancing across the globe, sometimes coming together, and sometimes splitting apart. The area of the global surface is limited, and inevitably, tectonic movements will push continents together or rip them apart. The Earth is a remarkable planet: our very existence as humans is inextricably linked with plate tectonics. When we look at other planetary bodies such as the Moon, Mars, Mercury, and Venus, we see different styles of tectonics, many with a 'fixed lid' structure (Moresi and Solomatov, 1995): the crusts of other Solar System terrestrial planets are mostly thick and rigid, and incapable of large horizontal motions (Venus may be an exception). In contrast, the Earth has a thin and more malleable upper layer (termed the lithosphere), some 280 km thick at most, with oceanic lithosphere being thinner, mostly varying between 50 and 140 km (down to zero kilometre thickness at actively spreadiing mid ocean ridges).

The lithosphere is underlain by a deeper mantle that convects and moves the lithosphere around like turbulent seas breaking up sea ice. When continents are pushed apart,as can be seen, for example, in the Red Sea today, new oceans are created. Oceans are more ephemeral features on Earth compared to continents. Oceans are born and die over c. 200 Ma (million years) periods. The oldest part of our modern oceans today is around 180–200 Ma (with some oldest ocean floor relics within the Mediterranean Sea aged at c. 340 Ma, Granot, 2016) in contrast to the oldest continent, aged 3.8–4 billion years in terms of large areas of crust, although some minerals within Earth have been dated at 4.4 billion years (Wilde et al., 2001), close to the age of the Earth itself, 4.56 billion years old. One of the many continental configurations that have existed on Earth included Gondwanaland, which held together as a supercontinent for around 380 Ma. Around 180 Ma, Gondwanaland began to break up into a myriad of continental fragments, creating new oceans: for example, the modern South Atlantic Ocean formed around 140 Ma.

Figure 1.3 presents a simple graphic of the complex tectonics that have led to the creation of the Himalayas. India (note Greater India, which is more extensive than the current geographic extent of South Asia) detached itself from Gondwanaland around 130 Ma. It then moved northwards at a high tectonic speed of c. 20 cm a year, crossing the Tethys Ocean (Peirce, 1975). The Tethys Ocean was an Indian Ocean-sized entity, consumed (or *subducted*, as known in tectonics) as other continents moved and forced Tethys to dive back (be subducted) into the mantle. This is the fate of all oceans: after being 'born' and growing, they are ultimately subducted back into the mantle, from whence they once came. Between India and Eurasia were other continental fragments that had their own unique geological histories separate

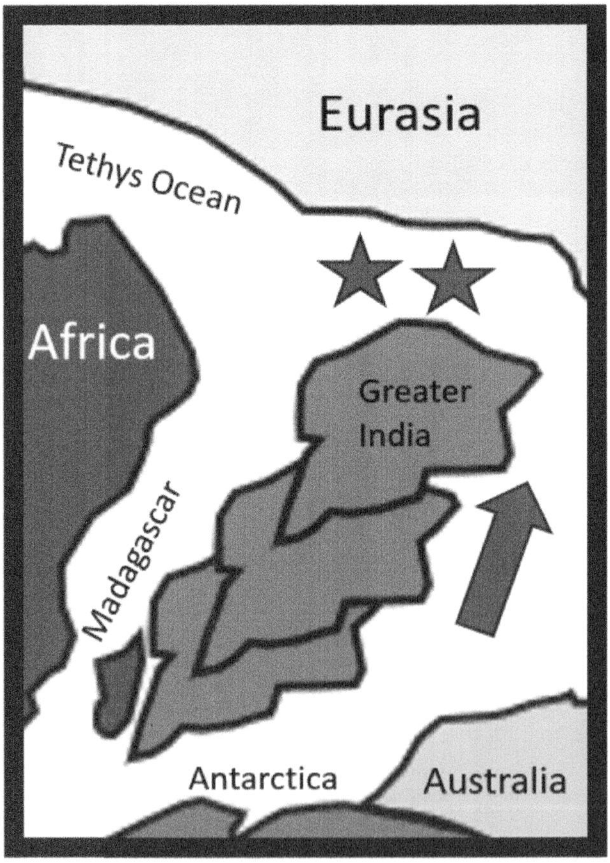

FIGURE 1.3 Northward drift of India. *Source*: Author's diagram. Note: Stars mark positions of other geological terranes that were caught up in the big collision: most notably the Kohistan-Ladakh and Karakoram terranes. North is upwards in this diagram.

from Eurasia and India. They became amalgamated within the larger collision, like small cars caught up between a head-on collision between two juggernauts on the road. Geologists debate how exactly the sequence and order of events happened, but the final result is there for all to study: we have a collage of geological terranes like pieces in a jigsaw that now form a continuous piece of crust (Treloar and Searle, 2019). Greater India began to collide with Eurasia and other crustal terranes from around 50 Ma. The rate at which India moved northwards slowed to c. 5 cm per year once collision began. The collision is ongoing with Greater India now 'underplated' (and underplating) beneath Tibet: this means it dived under Eurasia, but because it is tectonically buoyant it cannot subduct (as oceanic plates do) and, instead, moves horizontally beneath Tibet and Eurasia, directly under the upper plate, thickening the crust to up to 80 km beneath Tibet. The average continental crust is c. 30–35 km thick. The next stage in understanding is how precisely the continental collision manifested itself and ended up making mountains and a large continental plateaus.

CONTINENTAL COLLISION AND MOUNTAIN MAKING

Once plates collide, a number of mechanical processes take over, manifesting in a myriad ways and culminating in the production of mountains: a tectonic process *termed orogenesis.* The momentum of the Indian Plate underthrusting Eurasia and other terranes from c. 50 Ma provides the colossal amount of energy required for orogenesis. The sheer size, volume, and mass of the whole Indian Plate colliding with Eurasia generates a titanic dynamic momentum (momentum = mass × velocity): consider the gargantuan mass of the Indian Plate and the kinematic response from the Eurasian Plate. Figure 1.4 illustrates a cross-section through the Indian and Eurasian Plates across Nepal and Tibet. The Indian Plate 'subducts' beneath Eurasia in a shallow manner, forming a moving floor to the Eurasian Plate (Treloar and Searle, 2019). The upper parts of the subducting Indian Plate 'scrapes' away and detaches from deeper parts in response to the coupling of India and Eurasia. Deformation is expressed in numerous forms, but perhaps the most significant mountain-forming deformation-process creates *thrust faults*. Geological faults are planes along which rocks are fractured and moved around. In the case of the Himalayas, the most

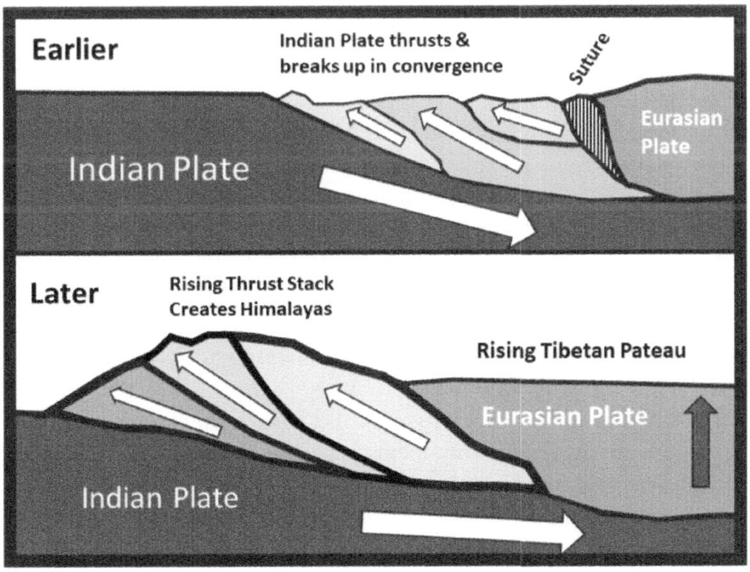

FIGURE 1.4 Tectonics of India underthrusting Eurasia. *Source*: Author's diagram.

common (thrust) structure responds to compression as the plates collide with one another. Rock masses are shortened and respond by creating faults at shallow angles that start to pile on top of one another and thicken the crust. The crust is simultaneously shortened and thickened under considerable deformation, generating abundant large magnitude earthquakes (Feldl and Bilham, 2006). The underplating of India beneath Eurasia doubles the crustal thickness and causes the Tibetan Plateau to rise. The boundary between two geological terranes, once separate but now joined, is termed a 'suture' (Treloar and Searle, 2019). A *suture* marks the final site of the destruction of the once intervening oceans, and forms the boundary between now cojoined but once separate plates. Intervening deep ocean floor material is pushed up and jammed between plates during the final collisional moments of plates: a journey that may take ocean volcanic rocks and deep-sea sedimentary rock from around 4 km beneath sea level to 1–8 km above sea level. As rocks are placed under enormous stress, they are heated up and buried to depths of tens of km; their engineering properties (*termed rheology*) change.

At 'normal' temperatures and pressures at the Earth's surface, rocks are brittle and break and snap through the action of geological faults. As they heat up, they eventually become ductile and respond to pressure in different ways, producing a wide range of ductile structures such as folds (strata that form 'wave-like structures') and shears (ductile 'faults' which move crust around at deep and hot levels). Once rocks are above around 400°C and pressures equivalent to around 2–4000 atmospheres, they can deform in a malleable ductile manner. Similarly, frictional heat along faults and within the vicinity of faults causes rocks to become increasingly ductile. Figure 1.5 presents a sand experiment, in this case undertaken at the Australian National University, Canberra, which beautifully illustrates how thrust stacks form. A transparent sandbox is filled with different coloured layers of sand representing horizontal layers of strata. The sand layers are then pushed from one side of the sand box via a tightening screw, representing tectonic collision. As the sand pile shortens, it thickens and produces a series of thrust faults which pile one on top of the other. There is a time sequence to the thrust faults: the youngest fault is the lowermost structure (No 4 in Figure 1.5) and the oldest (first formed) structure becomes the most highly elevated thrust unit (No 1 in Figure 1.5). Alongside thrusting, the sand stack begins to form a fold structure: in this case, an upturned U shape beginning to form. As thrust stacks grow and thicken, the heat and pressure of the system increases, moving the rheology of parts of the pile from the brittle to the ductile state. Radiogenic elements within the thrust pile add to the heating budget. Eventually, temperatures can reach such high levels, from 650°C to c. 1000°C, causing rocks to melt, forming liquid magmas that can stay within the rock mass and lubricate the deformation, or collect in large volumes, creating balloon-shaped volumes of granitic magma which rise and freeze at different layers in the crust. All these tectonic orogenic processes accumulate to increase crustal mobility and allow crust to move around as directed by tectonics: in this case, to create an elevated plateau and the Himalayan Mountain range.

The whole Asian continental dimension of the Himalayan tectonic story is termed *indentation tectonics*. Two pioneering French geophysicists, named Molnar and Tapponnier, explained succinctly how the Asian plate has been extensively deformed by India 'indenting' into Eurasia (Molnar and Tapponnier, 1975 and Tapponnier et al., 1982). Figure 1.6 summarises the models of visionary Molnar and Tapponnier. The experiment on the left-hand side of Figure 1.6 presents the results of striped plasticine being pushed (indented) by a piston. The piston represents the indenting Indian Plate, and the plasticine represents the indented Eurasian Plate. The experiment demonstrates shortening and compression in front of the piston, but also reveals a whole spectrum of deformation patterns further away from the immediate site of collision. Faults appear at all sorts of angles, with plasticine moving horizontally past each other: these are termed *strike-slip faults* that allow crustal blocks to move to the left or right of adjacent crustal blocks, depending on the local stress regime. The cartoon on the right-hand side of Figure 1.6 maps black lines (faults) with red and yellow arrows indicating local crustal movements in Asia. India itself is bound by eastern and western fault sets that allow it to move northwards. The arcuate red fault in northernmost India marks the point of the main tectonic collision at the present time, south of the Himalayas. Black faults in and around Tibet move eastwards, pushing parts of the Tibetan crust towards the East. North of Tibet,

FIGURE 1.5 Thrusts and sandboxes. Note: Stills taken from a video that records how sand is deformed as it is shortened, demonstrating propagation of thrust faults. *Source*: Acknowledgements to Stephen Cox and ANU University, Canberra; https://www.youtube.com/watch?v=KvYLuOdf4AY.

faults move in a strike-slip manner but also extend the crust, creating cracks and rift valleys: one of which is filled with water to create Lake Baikal in Siberian Russia. Faults in East China, Myanmar, Malaysia, Thailand, and Sumatra all move crustal units, in a relative sense, to the southeast. Molnar and Tapponnier, from their own work and through the collation of other researchers' work, mapped out the complex fault pattern of Asia in the 1970s and worked out the relative movements of every fault. This inspired their *Indentation Tectonics Model*, which has inspired tectonicists ever since. Latest high-resolution satellite GPS data, which has provided quantitative data for the vectors of crustal movement since the 1990s, verifies and adds great value to the Indentation Tectonics model (ESA 2026). The Indentation Tectonics model has revolutionised geologists' understanding of continent-continent collisions and inspired many researchers to dig deeper and think more boldly about continental architectures related to tectonic collisions. New data and models increasingly refine and develop indentation tectonics thinking, and crustal and mantle deformation in the Himalaya-Tibet region.

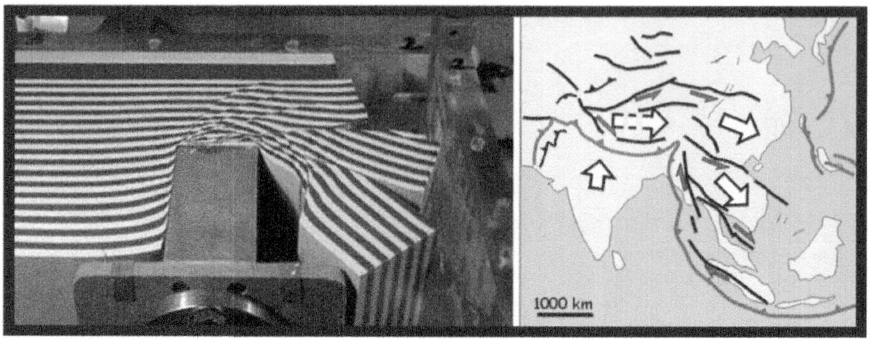

FIGURE 1.6 Indentation tectonic model for the India Eurasia collision. *Source*: Acknowledgements to Frasca et al., 2021. European Geological Union Blog https://blogs.egu.eu/divisions/ts/2021/02/15/ts-must-read-tapponnier-et-al-1982-propagating-extrusion-tectonics-in-asia. Reproduction allowed under Creative Commons licence CC by 4 international attribution. https://creativecommons.org/licenses/by/4.0/.

REFERENCES

Clamosa, F. (2018). View of Gondwana at 420Ma from a South Polar Projection. https://commons.wikimedia.org/wiki/File:Gondwana_420_Ma.png.

Feldl., N., & Bilham, R. (2006). Great Himalayan earthquakes and the Tibetan plateau. *Nature,* 444(7116), 165–170.

Frasca, G., et al. (2021). European Geological Union Blog. https://blogs.egu.eu/divisions/ts/2021/02/15/ts-must-read-tapponnier-et-al-1982-propagating-extrusion-tectonics-in-asia.

Granot, R. (2016). Palaeozoic oceanic crust preserved beneath the eastern Mediterranean. *Nature Geoscience*, 9(9), 701–705. https://doi.org/10.1038/ngeo2784.

Hedin, S. (1909). *Trans-Himalaya*. Volume 1. London: MacMillan Publications, 949pp.

Hergé. (1962) [1960]. *Tintin in Tibet*. Translated by Leslie Lonsdale-Cooper and Michael Turner. London: Egmont.

Hilton, J. (1933). *Lost Horizons*. London: MacMillan Publications, 270pp. ISBN 978 1840243636.

Molnar, P., & Tapponnier, P. (1975). Cenozoic tectonics of Asia: Effects of a continental collision. *Science*, 189, 419–426. https://doi.org/10.1126/science.189.4201.419.

Moresi, L.-N., & Solomatov, V. S. (1995). Numerical investigation of 2D convection with extremely large viscosity variations. *Physics of Fluids*, 7(9), 2154–2162.

Peirce, J. W. (1975). Palaeomagnetic evidence from DSDP cores of northward drift of India. *Nature,* 257, 570–572.

Suess, E. (1833). *Das Anlitz der Erde*. Published in volumes between 1883 and 1909. Prague, Vienna, Leipzig: F. Tempsky Publications, 2954pp.

Tapponnier, P., Peltzer, G., Le Dain, A. Y., Armijo, R., & Cobbold, P. (1982). Propagating extrusion tectonics in Asia: New insights from simple experiments with plasticine. *Geology,* 10(12), 611–616.

Treloar, P. J., & Searle, M. P. (Eds.). (2019). *Himalayan Tectonics: A Modern Synthesis*. Geological Society, London, Special Publications, 483.

Wilde, S. A., Valley, J. W., Peck, W. H., & Graham, C. M. (2001). Evidence from detrital zircons for the existence of continental crust and oceans on the Earth 4.4 Gyr ago. *Nature,* 409(6817), 175–178.

Younghusband, F. E. (1910). *India and Tibet*. London: John Murray Publications, 455pp.

Geological Framework and Terrane Collage of Himalaya-Tibet 2

Controls on Asian River Drainage

As geologists developed ever more sophisticated geological and structural maps of the Himalaya-Tibet region, it became increasingly apparent that continental collision between the mega-plates of India and Eurasia also involved the interaction of numerous smaller plates (Gansser, 1964; Treloar and Searle, 2019; Petterson, 2010, 2019, 2023; and references therein). Geologists defined a series of *geological terranes* within the Himalayan-Tibet region: a geological terrane is defined as a discrete crustal unit with a unique, coherent geological history and structure, separated from other terranes by geological fault structures. The accretion of terranes involved in the Himalayan Orogeny, and during preceding geological times, produced a new geological framework comprised of a terrane collage. Some geological terranes were once island arcs (that formed and evolved within the Tethys Ocean) and were tectonically 'swept up' via the India-Eurasia collision. Other terranes had more complex histories and were accreted to Eurasia during pre-Himalayan Orogenic times. Whilst the term 'final docking' of the Indian and Eurasian Plates is useful from a conceptual point of view, the reality is that the Himalayan region remains dynamic, with plates continually jostling with one another. Figure 2.1 depicts the giant >8,000 m asl(metres above sea level) Nanga Parbat Mountain in North Pakistan. This view presents one of the steepest continental slopes in the world. The difference in elevation between the lower Indus Plain area and the top of Nanga Parbat is over 5,000 m within a short horizontal distance. Nanga Parbat is a physical representation of the sheer dynamism of the Himalayas. It is the northernmost point of the Indian crust, surrounded by faults, and marks the *suture* (a geological structural boundary between two tectonic plates) separating the Indian and Kohistan-Ladakh plates. In plan view, the mountain is cigar-shaped, bounded by active faults moving the Indian crust upwards and northwards at around 1 cm per year (Zeitler, 1985; Butler, 2019).

Figure 2.2 displays a simplified geological and tectonic map of the Himalayas and Tibet. In essence, the map is divided into several zones and terranes, separated by major crustal fault structures. The Main Frontal Thrust structure separates the Lesser Himalaya, lower in general altitude, from the Indo-Gangetic Plain. This zone is characterised by a range of tree-covered mountains that rise in elevation between low-lying hills and ridges up to 4,500 m asl. The Main Central Thrust separates the Lesser Himalaya from the High Himalaya with mountains rising to over 8,000 m asl, progressively becoming drier, icier, and devoid of significant vegetation cover as the watershed is passed, and the mountains move into the colder, thinner atmospheric layers. The longest and most significant plate boundary, the Indus-Tsangpo Suture Zone (Figure 2.2), extends 2,500 km from Afghanistan and Pakistan to NE India and Nepal/Myanmar.

The Indus-Tsangpo Suture Zone marks the northern junction of the Indian Plate with the Eurasian Plate, which in turn is comprised of a number of terranes, such as the Lhasa and Qiangtang terranes (Figure 2.2). In the northwest, there are terranes which wrap around the NW promontory of the diamond-shaped Indian continent: the Kohistan-Ladakh and Karakoram Terranes, which are separated by the Shyok Suture Zone, in turn faulted by a major structure named the Main Karakoram Fault (Treloar and

DOI: 10.1201/9781003794868-2

FIGURE 2.1 The 8,000 m + Nanga Parbat Mountain from the 3,000 m high Indus Valley near Jutial, North Pakistan. *Source*: Author's photograph.

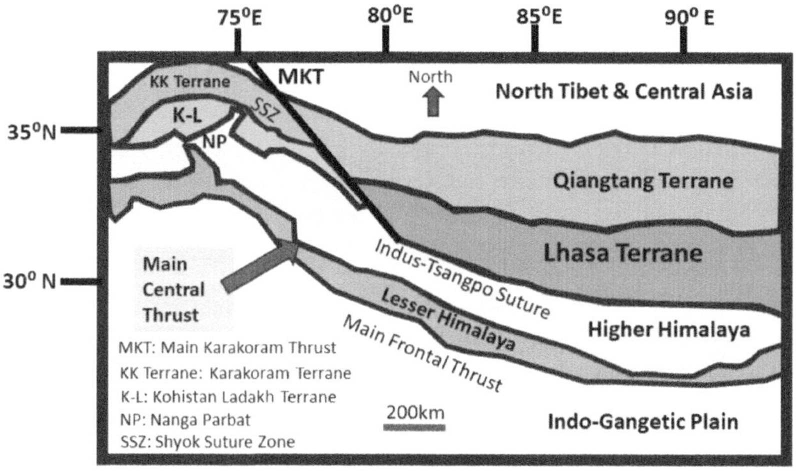

FIGURE 2.2 Simplified geo-tectonic map of the Himalaya-Tibet region. *Source*: Author's diagram.

Searle, 2019 and references therein). Structures such as the Main Boundary, Central, and Karakoram Thrusts, alongside the suture fault zones, remain variably active. The Main Boundary Thrust and faults around Nanga Parbat are examples of highly active faults that produce large earthquakes. The Main Central Thrust (MCT) was (and is) perhaps the largest and most important structure in terms of the sheer volume of crust mobilised and the degree of mountain-building the MCT was directly involved in. Earthquakes generated within continent-continent collisions can be particularly dangerous, with large magnitude earthquakes (6->8 on the Richter scale) forming at shallow depths (15-<5 km) and transmitting powerful energy waves to the surface, precipitating landslides, blocking rivers, and constantly reshaping landscapes (Bilham, 2019). Recent earthquakes in Kashmir, Pakistan, (2005), and Kathmandu, Nepal, (2015), are examples of earthquake-induced disasters that killed over 100,000 people and made millions homeless (Hussain et al., 2009; Hubbard et al., 2016). Advances made in Himalayan geotectonics have led to a deeper understanding of the causes and likely locations/magnitudes of future earthquakes:

this area remains research in progress with significant promise in delivering high-quality scientific advice for disaster risk reduction (Bilham, 2019).

HOW THE HIMALAYAS CHANGED CLIMATE AND CREATED MODERN GEOGRAPHIES

Figure 2.3 clearly illustrates the continental-wide impact of the Himalayan Mountain building event, not only in creating the biggest mountains and largest high plateau on Earth, but in shaping a profound new geography that changed the world, including human society. Modern drainage networks, much of which date back millions of years, formed as the Himalayas and Tibet rose, originating in both western and eastern Tibet (Ternes, 2019). Mount Kailas in western Tibet is the ultimate geographical source of the Indus, Sutlej, Ganges, Brahmaputra, and Yarlung-Tsangpo rivers. Mount Kailas is worshipped as a divine mountain by Hindu, Buddhist, Bon, and Jain followers, who recognise the profound importance of the great mountain. Eastern Tibet is a source for the Yellow, Yangtze, Mekong, Salween, and Irrawaddy rivers. Tibet, often known as the 'water tower of the Asian world', sources rivers that have produced fertile agricultural plains in Pakistan, Punjab, North India, Nepal, Bangladesh, Myanmar, Thailand, Cambodia, Laos, and eastern China. Billions of people have arisen over time, partly or largely through relative food abundance, watered by continental-scale river systems. As mountains and the Tibetan Plateau arose, they changed atmospheric pressure systems and the movements of seasonal rains or monsoons. Debate revolves around when exactly the mountains and Tibet achieved the heights we see today, but it appears likely that Tibet became a high plateau from as late as 14 Ma ago; some argue for earlier ages. The Main Central Thrust was particularly active from around 25 to 16 Ma, which created the High Himalayas south of Tibet (Searle et al., 2024). The rise of Tibet as a monolithic entity is hotly debated within the literature, with some arguing the origins of the plateau extend back to pre-Himalayan Orogenic times. Alternate models argue that the creation of key elements of the Tibetan Plateau occurred sequentially from Mesozoic times, although the final formation of a large intra-continental plateau with an average height over 5,000 m asl probably occurred from c. 20 Ma (Spicer et al., 2021).

The Himalayan-Tibet topographic barrier became highly influential in changing rainfall and weather patterns (Figure 2.4). Air pressure and humidity differentials between central Asia and south, east, and

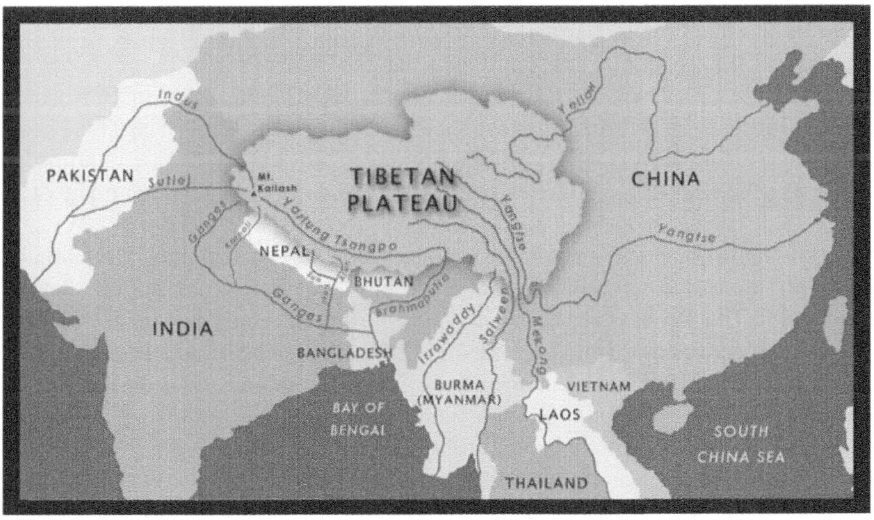

FIGURE 2.3 Map of Major Asian rivers owing their source to Tibet. *Source*: Acknowledgements Creative Commons: https://commons.wikimedia.org/wiki/File:Rain shadow effect.jpg. Meg Stewart. Issued using Creative Commons licence CC by 2 international attribution. https://creativecommons.org/licenses/by-sa/2.0/deed.en

FIGURE 2.4 Orographic Rainfall. Note: As clouds rise over mountain barriers, they cool and rain or snow. Once they pass the high mountains, they have lost the bulk of their moisture, and the interiors are much drier. *Source*: Acknowledgements Creative Commons: https://commons.wikimedia.org/wiki/File:Rain_shadow_effect.jpg. Meg Stewart. Issued using Creative Commons licence CC by 2 international attribution. https://creativecommons.org/licenses/by-sa/2.0/deed.en

southeast Asia shape the nature, intensity, duration, and temporal periods of Asian Monsoon rainfall. The phenomenon of orographic rainfall results in the southern sides of the Himalaya developing a wet forest ecosystem in response to clouds progressively precipitating rain and snow as they rise and lose their ability to retain water vapour (Figure 2.4). Once clouds reach the High Himalaya and lands beyond, they have either disappeared or lost the great bulk of their humidity. Tibet and the Trans-Himalaya north of the high mountain watersheds are mountain deserts, unable to support forests or extensive tracts of vegetation. Inner Asia also dried out as the Himalayas matured: much of Mongolia and its surrounding regions are relatively parched or desert. Prior to the Himalayan collision, the world had been in a much warmer period for hundreds of millions of years. During the dinosaur and reptile-dominated Mesozoic Period, the world was perhaps up to 10–14ºC warmer than the pre-industrial average global temperatures (Hallam, 1985). This world was an ice-free world. Following the rise of the Himalayas and the subsequent massive levels of erosion that locked up greenhouse gases in the sedimentary basins and deltas of the Gangetic Plain and Bangladesh, the earth cooled (Raymo and Ruddiman, 1992). By around 34 Ma, ice began to appear in large quantities, particularly in Antarctica, with the most recent and extensive major ice age beginning at 2 Ma. A cool Earth facilitated the accumulation and development of large glacier systems within the High Himalaya: the Baltoro glacial system of the Karakoram in Northern Pakistan is particularly extensive. This growth in Himalayan ice fields led to the term 'The Third Pole' on earth, as it is the largest ice reservoir beyond the North and South polar regions. Modern anthropogenic climate warming is reversing the Himalayan climate cooling effect and has begun to melt many of the Himalayan glaciers.

Mountain systems strongly influence the development of human societies. Mountain barriers divide human groups who develop and evolve in isolation from other groups, creating unique cultures. *Homo sapiens* first arrived in Tibet from at least 20,000 years ago, some argue a date of 40,000 years ago (Zhang et al., 2018). Petroglyphs (rock art) from Ladakh and Zanskar Himalaya indicate more permanent human habitation for at least 11–14,000 years, with indications of humanoids back to the Palaeolithic period (Deshpande, 2024). Skulls discovered in Mongolia date back to c. 35,000 years ago (Deviese et al., 2019). The oldest South Asian sapiens remains come from Sri Lanka from c. 45,000 years ago (Wedage et al., 2020). China has revealed evidence of the earliest sapiens occupation from around 80–120,000 years ago (Liu et al., 2015). Whilst mountains form barriers, humans find routes to trade along and through, locating weak points in the barriers. The numerous Silk Routes through and around the Himalayas and central Asia are particularly famous. Cross-mountain roads that hosted yaks, horses, people, and goods for centuries allowed people to mix and influence one another. Silk Routes are predominantly east-west or northwest-southeast, following the 'grain' of the mountains, or river channels of the mountains and the most accessible lower mountain passes. Invasions of later groups into the Greater Himalayas, Tibet, Afghanistan, India, and eastern-south-eastern Asia led to power struggles, wars, battles, confrontations, and genetic mixing. Central Asian and Turkestan powers exerted influence for centuries in modern Afghanistan, Pakistan, North India, and parts of the Trans Himalaya. Tibetan people extended their influence across much of the Himalayan region and had close alliances with Mongolia. China sought control over Tibet for centuries, but Tibet successfully achieved isolation and independence for the bulk of its history until recent invasions of Maoist China. Modern India offered refuge for Tibetan people across much of India, including within Ladakh and Himachal Pradesh. The now hard-border zones between China and India were porous for generations. South and western Tibetans freely mixed with their counterparts in Arunachal Pradesh, Nepal, and the NW Trans-Himalaya. Trade routes along the trend of the Trans-Himalaya were active in pre-Indian continent partition times: Nepalis, Tibetans, Ladakhis, Baltis, Gilgitis, Chitralis, and Hunza people traded and shared ideas, technologies, crafts, and cultures. Whilst mountains create formidable physical barriers and help shape the character of humans across Asia, humans also find ways through the mountains as needs and curiosities drive them. Hardening of geo-political boundaries runs counter to the flow of cross-Himalayan human mixing histories, but may become more permeable with time.

REFERENCES

Bilham, R. (2019). Himalayan earthquakes: A review of historical seismicity and early 21st century slip potential. *Geological Society, London, Special Publications*, 483(1), 423–482.

Butler, R. W. H. (2019). Tectonic evolution of the Himalayan syntaxes: The view from Nanga Parbat. In P. J. Treloar & M. P. Searle (Eds.), *Himalayan Tectonics: A Modern Synthesis*. Geological Society, London, Special Publications, 483, 215–254. https://doi.org/10.1144/SP483.7.

Deviese, T., Massilani, D., Yi, S., Comer, K., Fewlass, H., Pantzuri, A., et al. (2019). Compound-specific radiocarbon dating and mitochondrial DNA analysis of the Pleistocene hominin from Salkhit, Mongolia. *Nature Communications*, 10(1), 274.

Deshpande, A. (2024). *Speaking Stones: Rock Art of Ladakh*. Hyderabad: Printed by Pragati Offset Pvt Ltd.

Gansser, A. (1964). *Geology of the Himalayas*. London: Wiley Interscience.

Hallam, A. (1985). A review of Mesozoic climates. *Journal of the Geological Society*, 142(3), 433–445.

Hubbard, J., Almeida, R., Foster, A., Sapkota, S. N., Bürgi, P., & Tapponnier, P. (2016). Structural segmentation controlled the 2015 Mw 7.8 Gorkha earthquake rupture in Nepal. *Geology*, 44(8), 639–642. https://doi.org/10.1130/G38077.1.

Hussain, A., Yeats, R. S., & Lisa, M. (2009). Geological setting of the 8 October 2005 Kashmir earthquake. *Journal of Seismology*, 13, 315–325. https://doi.org/10.1007/s10950-008-9101-7.

Liu, W., Martinón-Torres, M., Cai, Y.-J., et al. (2015). The earliest unequivocally modern humans in southern China. *Nature*, 526, 696–699. https://doi.org/10.1038/nature15696.

Petterson, M. G. (2010). A review of the geology and tectonics of the Kohistan island arc, north Pakistan. *Journal of the Geological Society of London, Special Publication*, 338, 287–327. https://doi.org/10.5194/essd-14-4171.

Petterson, M. G. (2019). Plutonic and volcanic rocks of Kohistan, North Pakistan: Implications for island arc genesis and evolution. In P. J. Treloar & M. P. Searle (Eds.), *The Himalayas: A Modern Synthesis*. Special Publication of the Geological Society of London, No 483, pp. 123–164. https://doi.org/10.1144/SP483.4.

Petterson, M. G. (2023). *Himalayan Thick-Skin Basement Deformation of the Ladakh Batholith, Leh-Ladakh Region, NW India*. Switzerland: Springer Nature (Geology) Publications.

Raymo, M., & Ruddiman, W. (1992). Tectonic forcing of late Cenozoic climate. *Nature*, 359, 117–122. https://doi.org/10.1038/359117a0.

Searle, M. P. (1986). Structural evolution and sequence of thrusting in the High Himalayan, Tibetan-Tethys and Indus suture zones of Zanskar and Ladakh, Western Himalaya. *Journal of Structural Geology*, 8, 923–936. https://doi.org/10.1016/0191-8141(86)90037-4.

Searle, M. P., Larson, K. P., & Stipska, P. (2024). The making of Mt Everest: Channel flow and low-angle normal faults. *Journal of the Geological Society*, jgs2024-118. https://doi.org/10.1144/jgs2024-118.

Spicer, R. A., Su, T., Valdes, P. J., et al. (2021). Why 'the uplift of the Tibetan Plateau' is a myth. *National Science Review*, 8(1), nwaa091. https://doi.org/10.1093/nsr/nwaa091.

Ternes, B. (2019). Exploiting Shangri-La: Assessing the Tibetan Plateau's natural resources and the work of Karl Wittfogel. *Review of European Studies*, 12, 1. https://doi.org/10.5539/res.v12n1p1.

Treloar, P. J., & Searle, M. P. (Eds.). (2019). *Himalayan Tectonics: A Modern Synthesis. Geological Society, London, Special Publications*, 483, 215–254. https://doi.org/10.1144/SP483.7.

Wedage, O., Roberts, O., Faulkner, P., Crowther, A., Douka, K., Picin, A., Blinkhorn, J., Deraniyagala, S., Boivin, N., Petraglia, M., & Amano, N. (2020). Late pleistocene to early-holocene rainforest foraging in Sri Lanka: Multidisciplinary analysis at Kitulgala Beli-Lena. *Quaternary Science Reviews*, 231, 106200.

Zeitler, P. K. (1985). Cooling history of the NW Himalaya, Pakistan. *Tectonics*, 4(1), 127–151. https://doi.org/10.1029/TC004i001p00127.

Zhang, X. L., et al. (2018). The earliest human occupation of the high-altitude Tibetan Plateau 40 thousand to 30 thousand years ago. *Science*, 362(6418), 1049–1051.

Regional Geology and Physical Geography of Ladakh

3

The geological foundations of any land, region, or country shape its territory, economy, and people. Solid rock substrate forms the literal foundations of any landscape: dig a little downwards through the plants, soils, and weathered rock or regolith, and you will quickly reach hard, solid stone. Different types of geology, or different rock types (termed *lithology* from *lithos*, an Ancient Greek language word), form different landscapes and terrains. Limestones, for example, being soluble, are hosts to caves and sinkholes. Mudstones and clays form more subdued or rolling topography. Granites and crystalline rocks can support steep crags and cliffs. Rocks are pushed, pulled, ripped apart, and forced together by tectonic forces. These forces produce geological structures such as faults and folds, which, in turn, shape landscapes. Erosive forces such as ice, water, and wind will hone in on and sculpt weaker parts of the geological foundations, such as faults, fold structures, and less cohesive lithologies. Figure 3.1 presents contorted bedded sedimentary rocks around Rangdum, Zanskar, which have been hardened through the Himalayan Orogeny. The bedded and contorted character of the rocks shapes the mountain, producing sharp, jagged outlines and serrated mountain edges. In a mountain desert climate, wind winnows, sand-blasts, and erodes rocks into manifold shapes, producing ledges and steep scarps, blowing away eroded surfaces. Deposits of wind-blown sand carpet parts of the mountains. Intense rainfall transports large volumes of sediment downwards along steep mountain channels that fan out as the mountain meets the plain, forming alluvial fans. The foreground of Figure 2.1 gives a perspective displaying a wide intra-montane flat plain and a Gompa or Monastery, perched on a rocky cliff, rising above the plain. Mountains are not merely towers of rock and ice but also include a wide range of landscapes, including numerous forms of valleys and intra-montane flatter-lands, where most people live.

Geology provides the raw materials of life for most human societies – Building materials for houses, shelter, roads, railways, airports, and cities, metals for implements, cables, wires, infrastructure, and manufactured goods, and energy in many forms, from fossil to green and renewable. In Ladakh, construction minerals are widely used in traditional house-building and for religious expressions such as mani walls and stupas, for maintaining ancient people and animal tracks, and for defining field boundaries through dry stone walls. Today, minerals are used for numerous developments related to urbanisation, water transport, and a surge in road building. Jewels and rock salt have been traded for millennia, and gold has been extracted on a small scale for local social and sacred contexts.

Living in mountain environments entails existing alongside geohazards including earthquakes, landslides, and occasional high-impact floods. Ladakhis have developed complex traditional wisdom, based on centuries of experience within the mountains. They understand the limits and dangers of the world around them. Whilst they have learned to develop settlements in safe places, and construct walking tracks through mountains and across icy rivers, new ways and growing populations bring different dimensions. Modern geological knowledge provides data on a history of hazard events such as floods and landslides, which leave their deposits in the rock record. Increasing levels of tectonic understanding assist in reducing seismic risk. A combination of old and new wisdom improves lives for all.

DOI: 10.1201/9781003794868-3

FIGURE 3.1 A well-exposed mountain in Rangdum, Zanskar, exposing contorted bedded strata. Mountains such as this shape the physical landscape of Ladakh. *Source*: Author's photograph.

REGIONAL TECTONICS AND GEOLOGY

Figure 3.2 presents the main geo-tectonic elements of the regions geographically close to Ladakh. Chapters 1 and 2 discuss tectonics in more detail. This chapter discusses those specific tectonic elements that impact the gross form of mountains and landscape. Three terranes are presented in Figure 3.2: the Indian Plate to the south, the Karakoram Terrane to the north, and the Kohistan-Ladakh Terrane sandwiched between the two. Kohistan-Ladakh is separated from the Indian Plate to the south by the Indus Suture, and the Karakoram Plate to the north by the Shyok Suture and the Karakoram Fault (Petterson, 2010, 2019, 2023). The terranes, situated at the apex of the NW Greater Himalayas, form an arcuate morphology, arching northwards. Ladakh itself and the SE Karakoram terrane are situated on the eastern side with a predominant NW-SE trend locally. Tectonics and major crustal structures, such as the sutures between terranes, are the strongest first-order control on the shape and orientations of landscapes.

Figure 3.3 presents a crustal scale geo-tectonic section, from SW to NE, beginning in the Indian Plate and heading across the Ladakh and Karakoram Terranes (Searle, 1983, 1986, 2024).

India underplates the terranes and Eurasia, doubling regional crustal thickness and generating the bulk of the stress that creates the mountains. The Indian Plate itself buckles and piles up through the stacking of southward-moving thrust faults such as the Main Central and Main Boundary Thrusts, forming the High Himalayan and Lesser Himalayan Ranges respectively. The Ladakh and Karakoram Terranes are deformed through the Himalayan collision, as well as having a history of pre-Himalayan collisions, as they themselves collided with Asia to the north (Treloar and Searle, 2019). This tectonic architecture in the third dimension (space and depth), as well as the fourth dimension (time: how deformation has progressed over millions of years), sets the deep foundation for the understanding of the mountains.

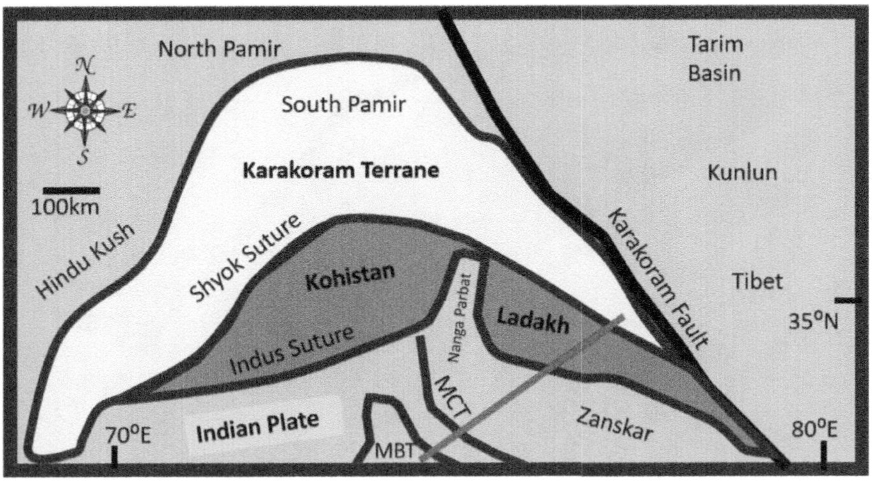

FIGURE 3.2 Tectonic terranes of the NW Greater Himalaya region. *Source*: Author's diagram. Note: Blue line indicates the position of the cross section shown in Figure 3.3. MBT: Main Boundary Thrust; MCT: Main Central Thrust.

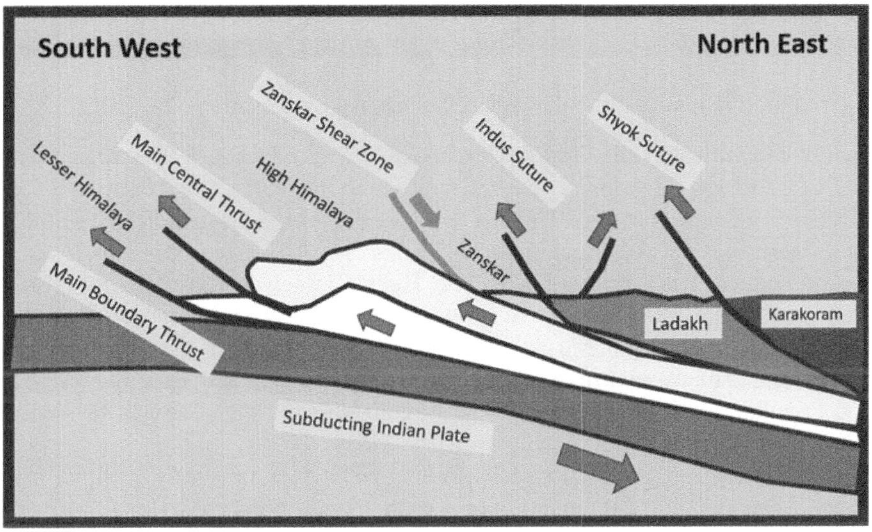

FIGURE 3.3 A SW-NE tectonic cross section through the Indian, Kohistan-Ladakh, and Karakoram terranes. Note: See Figure 3.2 for the cross-section location.

Figure 3.4 sets out the geological framework of Ladakh and will be referred to regularly as we 'travel' through the different elements of the geology of Ladakh from the southwest to the northeast (Searle, 1983). As was noted above, during the discussion of the tectonic framework, the geological units of Ladakh broadly extend from the northwest to the southeast. This is called the regional *strike* direction. The oldest rocks in the region are the Greater Himalayan Crystalline Unit, part of the Indian Plate. Some rocks from this unit date back to almost 2 billion years, from the Proterozoic Period of Earth history. The full rock spectrum of the Indian Plate sequence spans a wide range of geological time, with the youngest (cover-sedimentary) units being Eocene in age (c. 50Ma) (Searle, 1983, 1986).

The High Himalayan Crystalline rocks have been metamorphosed through heat and pressure, even up to the point of melting. The Tso Morari Complex is a fascinating high-grade metamorphic unit with an original Cambrian-Ordovician (480-450Ma) to Permian/Triassic (c. 250Ma) age. It has a later

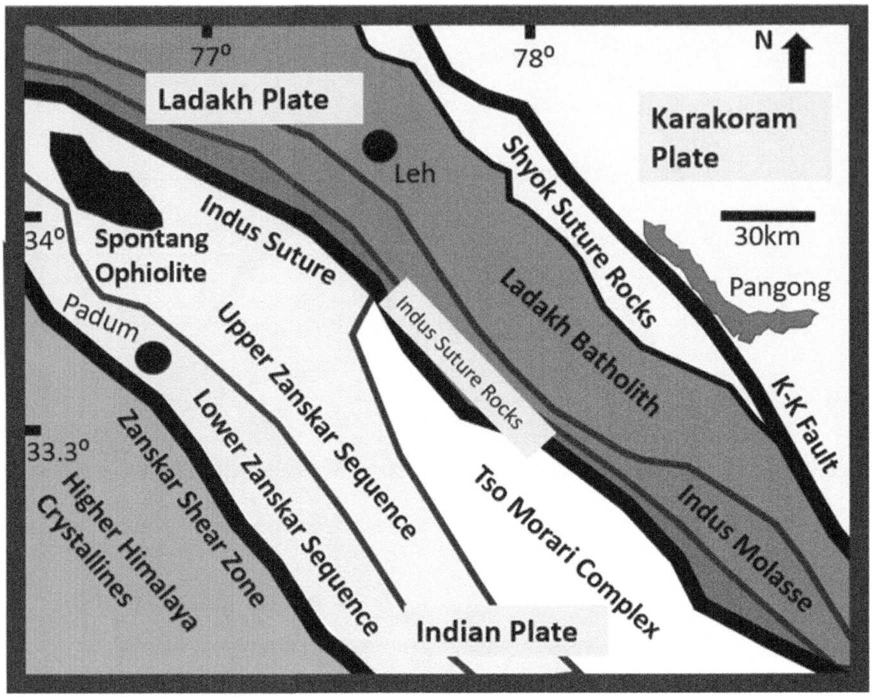

FIGURE 3.4 Geological map of the Ladakh region. *Source*: Author's map.

Himalayan-aged tectonic history that dragged the rocks down to depths of 90 km, as part of the down-plating Indian Plate, and was subsequently pushed upwards to the surface, now forming a domal structure: like a crustal 'bubble' (Bidgood et al., 2024). As much of the unit is formed of high-grade metamorphic gneiss rock, the structure is termed a 'gneiss dome'.

The Zanskar Sequence is separated from the High Himalayan Crystallines by a major crustal structure called the Zanskar Shear Zone. The lower part of the Zanskar Sequence is aged between Late Precambrian-Cambrian (c. c.550-485Ma) and Permian (c. 299-252Ma), comprising low-grade metamorphosed mudstones, sandstones, and limestones. The Upper Zanskar Sequence is mostly aged between the Lower-Mid Permian- and Lower Tertiary (c. 285–c.50 Ma) and includes well-bedded, highly folded sandstones, siltstones, and limestones (McElroy et al., 1990).

The Indus Suture separates the Zanskar, Tso Morari, and High Himalaya Crystallines from the Ladakh Plate (or Terrane). Along the suture, remnants of the ocean floor are preserved. Deep ocean floors form over 50% of the surface of the modern Earth. They are composed of volcanic rock called basalt, which forms from the melting of the Earth's mantle. Above the basalts are deep-sea sediments which are hard, flinty, and silica-rich, termed *chert*. Sutures mark the final collision and docking structures between two plates. The intervening ocean has been destroyed, but at the climax of continental collision, some remnants of the ocean floor are jammed between the colliding plates and obducted, or 'pushed up', onto the colliding subaerial continents. Slivers of basalt, chert, and deep ocean crustal rocks are located within suture zones. Sometimes complete *'obducted'* (tectonically thrust upwards onto continents) ocean sequences, *termed 'ophiolites'*, are found within the Indus and Shyok Suture zones. Mixtures of detached oceanic blocks, limestones, sandstones, and other lithologies, termed 'melange', are also located in suture zones. Indus Suture Rocks, located north of the suture, comprise metamorphosed and unmetamorphosed sedimentary rocks and volcanic units termed the Jurassic-Cretaceous aged Dras Volcanics (Petterson, 2019).

The 60–6 Ma Indus Molasse is a bedded sequence of sedimentary *conglomerates* (rocks which have a wide variation in particle size from boulder to mud), sandstones, siltstones, and mudstones. A *'molasse'* forms as mountains rise and sheds large volumes of sediment in the form of alluvial fans, which are further reworked by high-energy rivers, and quiet intra-montane lakes (Petterson, 2023 and references therein).

The 150–30 Ma Ladakh batholith is an amalgam of *intrusive igneous rocks*. An igneous rock is one which was once liquid magma and crystallised as it cooled and solidified. Some igneous magmas erupt at the Earth's surface, solidifying as volcanic rocks. The great bulk of magmas freeze at depth and form much of the continental crust, making the continental crust less dense and able to exist at the Earth's surface for perhaps billions of years. As the magmas rise within the crust because of their relative buoyancy, they force themselves into higher crustal levels, thus intruding into pre-existing rocks and are termed '*intrusive rocks*'. The Ladakh batholith (Petterson, 2023) comprises a wide range of intrusive rocks called gabbro, diorite, granodiorite, and granite (these terms will be explained in subsequent chapters).

The Shyok Suture Zone rocks, aged between c. 115–70 Ma, are similar in lithology and character to the afore-described Indus Suture rocks, as they have formed within analogous tectonic settings and include obducted ocean floor materials, ophiolites, and melanges, alongside volcanic units (Martin et al., 2025; Treloar and Searle, 2019).

The mostly Jurassic-Cretaceous aged Karakoram Plate, with some igneous activity progressing to much younger times (35–13 Ma), has many similar geological characteristics to Ladakh, including the large Karakoram batholith and a series of metamorphic, volcanic, and sedimentary rocks. The southernmost part of the Karakoram Terrane in northern Ladakh includes oceanic and continental volcanic rocks, a wide range of metamorphic rocks, and granites (Treloar and Searle, 2019 and references therein).

To summarise the geology in terms of what it represents tectonically helps the broad understanding of where each geological element fits in. The Greater Himalayan Crystalline, Tso Morari Complex, and the Zanskar geological units are part of the Indian Plate. The Indian Plate has a long history of geological evolution extending back across billions of years. When India broke free / became detached from Gondwana and drifted northwards, the northern edge of India became what is known as a *passive margin*. A passive margin is well illustrated today by the two sides of the modern North Atlantic. Western Europe and eastern North America are tectonically drifting apart from one another 'quietly', with limited present-day active tectonics. Hence, the margins are termed *passive*. When India collided with Kohistan-Ladakh/Asian Plate, the formerly passive margin then became an active continental collisional margin, which led to the deformation of the Greater Himalayan Crystalline, Tso Morari, and Zanskar geological units. The Kohistan-Ladakh Terrane was an island arc, caught between Asian and Indian Plates. Subduction of the Tethys Ocean beneath Kohistan-Ladakh produced a batholith and volcanism. Collision (of Kohistan-Ladakh) with the Karakoram Terrane/Asian margin at around 90 Ma (other tectonic models are presented in the literature, see Petterson, 2023 and references therein for further reading) created an *Andean Margin* on the souhern boundary of Kohistan-Ladakh. Andean margins are continental margins, under which oceanic plates subduct. The modern western side of South America experiences subduction from the Pacific and related oceanic plates, which have built the Andean Mountains (the second-highest mountain chain in the world). An Andean margin is an active margin that experiences earthquakes and large-volume magmatism. From around 50 Ma, Kohistan-Ladakh was then caught up within the Himalayan collision, as India finally collided with all the terranes of Asia. The Karakoram Terrane had a more complex history compared to Kohistan-Ladakh (discussed in Chapter 5), originating as part of continental Gondwanaland, drifting across Tethys Ocean from the Permian, and converted to an Andean margin following collision with with Asia during the Jurassic-Cretaceous, and later experienced collisions to the south from the Ladakh-Kohistan Island Arc, and India (Treloar and Searle, 2019).

GROSS ARCHITECTURE OF THE LADAKH MOUNTAINS

Figure 3.5 presents a summary of the geographic location of the main mountain belt components of north-western India and northern Pakistan. There are five sub-units of the Greater Himalayas: the Lesser and Greater Himalaya Ranges, the Zanskar Range, the Ladakh Range, and the Karakoram Mountains. The Lesser Himalaya forms the southern frontal ranges of the Himalaya between the plains to the south and the Greater Himalaya to the north. This range is also termed 'the Siwalik or Sivalik Range' (after Lord Shiva) and has an average elevation of 1,500–2,000 m high. The Greater Himalayas in this region include the >8,000 m Nanga Parbat and are bounded to the south by the Main Central Thrust and to the north by the Zanskar Shear Zone (Figure 3.4). The Zanskar Mountains are bounded to the south by the Zanskar Shear Zone and to the north by the Indus Suture Zone, rising to over 6,500 m. The mountains contain a number of lower-altitude ranges, deeply cut by river gorges. The Ladakh Range is dominated by the dissected Ladakh Batholith to the north and the Indus Molasse mountains to the south. Highest peaks are >6,000 m with valleys mostly at heights of around 3,500 m. The Karakoram Mountains are the second-highest range in the whole Himalaya-Tibet region, including >8,000 m peaks such as K2 (world's second-highest mountain), over 100 peaks over 7,000 m, and the large Baltoro Glacier network.

The major rivers of Ladakh are shown in Figure 3.6. These rivers include the mighty River Indus (twenty-first longest river in the world) and rivers that drain into the Indus, including the Zanskar, Kargil, Shyok, Nubra, Suru, Tsarap, and Stod Rivers. These rivers, in reality, form a single river network, or large-scale drainage system. They drain the mountain ranges of the Greater Himalaya, Zanskar, Ladakh, and Karakoram. River orientations are strongly controlled by first and second-order geological structures. First-order geological structures (see Figure 3.4) include the Zanskar Shear Zone and the Indus and Shyok Sutures. The NW-SE trend of these structures strongly influences the major orientations of Ladakh rivers, which stretch in a broadly NW-SE direction (e.g., the Indus, Shyok, Tsarap, and Stod Rivers). The second most common river orientation (NE-SW) is at right angles to NW-SE structures. Rivers such as the Zanskar, Suru, and parts of the Tsarap are broadly oriented NE-SW and reflect the second-order (fault) structures that form as a consequence of local tectonic vectors. Ladakh is a classic example of a *trellis-shaped river network* architecture (Marshak, 2019): networks characterised by rivers that run approximately parallel and perpendicular to each other. Fault-controlled young mountains tend to produce trellis river patterns. Other common river network architectures include radial and dendritic systems. *Radial river networks*, like the spokes of a wheel, are produced around high peaks separated widely from other peaks, such as high volcanoes. *Dendritic patterns* are possibly the most common river network, named after tree geometries, which they resemble. Dendritic networks form within a range of landscapes, including lowland and older, worn-down mountain terrains. Rivers drain *catchments,* which define the area around a river that delivers water to the river and the subterranean geology. Catchments are divided from one another by '*watersheds*', like the sloping roof of a house. All terrains divide up into catchments and watersheds, and the boundaries of catchments (watersheds) decide which way the water flows: one side or the other of the watershed (Marshak, 2019). Mountain Passes exist on such watersheds and include

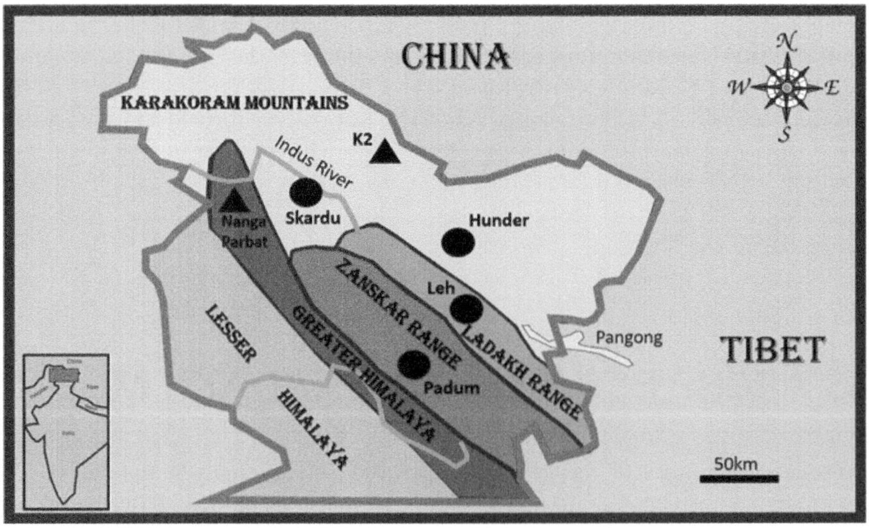

FIGURE 3.5 Mountain ranges of Ladakh. *Source*: Author's diagram.

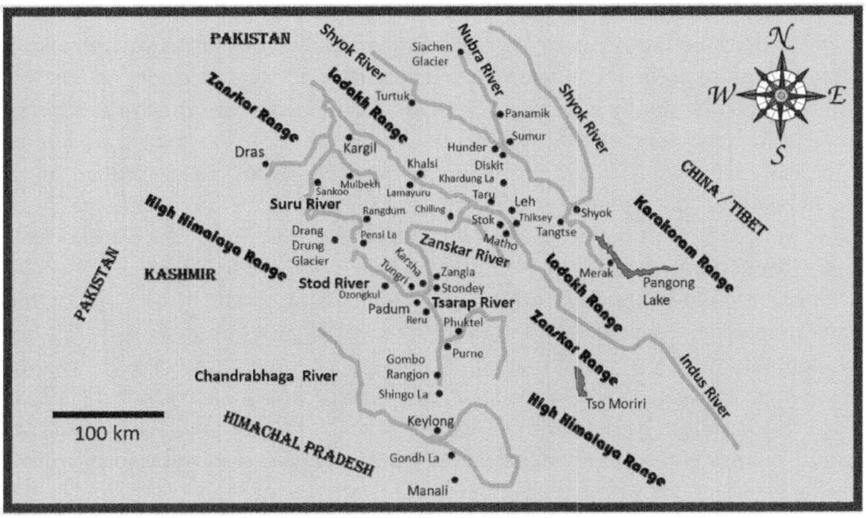

FIGURE 3.6 Major rivers, lakes, towns, villages, and settlements of Ladakh. *Source*: Author's map.

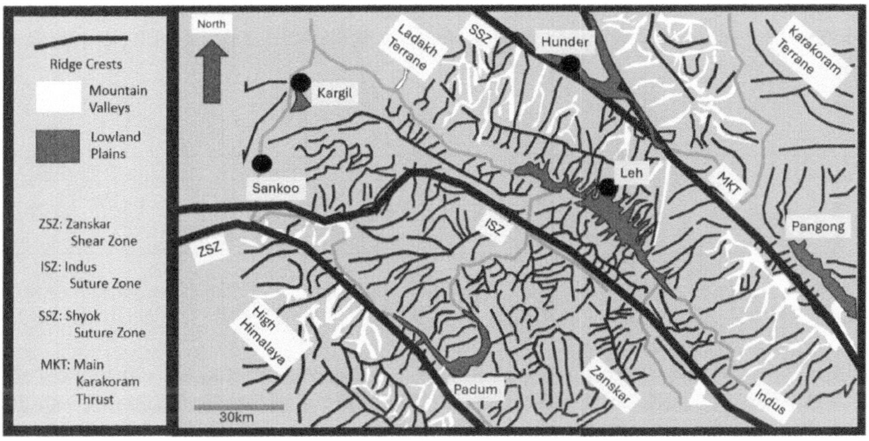

FIGURE 3.7 Mountain ridge crest and valley lines as indicators of mountain architecture and controlling geological structures. *Source*: Author's map.

Shinku-La, Pensi-La, and Khardung-La within Ladakh. There are numerous mountain passes in Ladakh linking valleys and communities, across which walking routes have passed for thousands of years. The very name of Ladakh means '*the land of mountain passes*'.

Figure 3.7 presents the results of an analysis of satellite images of Ladakh. Ridge crests (many of which are also watersheds), mountain valleys, lowland valleys, and rivers are presented alongside major geological structures. These data bring additional detailed insights into the architecture of the Greater Himalaya. Zanskar, Ladakh, and Karakoram mountain ranges. The Greater Himalayas form longer, NW-SE trending mountain ranges, with shorter, subordinate NE-SW ridges and mountain valley systems. The Greater Himalaya mountains include a significant number of older glacial valleys and presently actively glaciated valleys, many oriented NE-SW. The Zanskar Shear Zone and the Main Central Thrust (further south, not shown in Figure 2.7) strongly control the orientation of the Greater Himalayan Mountain Range features in Ladakh. The geometry of the Zanskar Range between the Zanskar Shear Zone and the Indus Suture presents a more complex story. Individual mountain ranges are shorter and exhibit a range of orientations with NE-SW as the majority orientation, and WSW-ESE, NNW-SSE, as secondary orientations. The Zanskar Shear Zone strongly influences the geometry of parts of the Stod and Tsarap rivers, as does the Indus Suture for rivers immediately south of the suture zone. The re-entrant (abrupt change in direction) in the Indus Suture towards Sankoo (change in direction locally from NW-SE to more N-S) influences topography and drainage locally. The greater variation in ridge crest orientations is a consequence of the presence of other structures within the Zanskar geology, such as the complex folding of much of the sequence. The smaller scale of mountain valleys is noticeable, suggesting a lower influence of glaciation on geomorphology (landscape 3-D shape) in the Zanskar area.

The Ladakh Range, located between the Indus and Shyok Sutures/Main Karakoram Thrust, has a strongly rectilinear/trellised form. Longest ridges are oriented NW-SE (with the exception of terrain SE of Leh, with predominant NE-SW trending ridges). Next-longest ridges and valleys are orientated NE-SW. Most valleys are oriented perpendicular to the River Indus, with an asymmetry in valley geometry to the NE and SW of the Indus. Valleys to the NE are up to 30km long, whilst those SW of the Indus extend only several kilometres. The Indus River has formed a relatively large and extensive valley plain centred on Leh. Larger glacially eroded valleys are much more extensive north of the Indus. Geological structures exert strong topographic control within the Karakoram Mountains. Lake, river valley, ridge crest orientations, and actively glaciated mountains within the Karakoram Terrane are strongly controlled by the Shyok Suture and Main Karakoram Fault (Figure 3.7). Similar to other mountains ranges examined here, the Karakoram range has longer NW-SE ridges and valleys and shorter NE-SW trending topographic elements.

To summarise, with local exceptions, the general and detailed architecture of the Ladakh mountain regions is dominantly shaped by tectonics and resultant geological structures. Deeper narratives of geology, structure, and physical geography of Ladakh, along with landscape development, are presented in subsequent book chapters.

REFERENCES

Bidgood, A., Parsons, A., Roberts, N., Waters, D., Tapster, S., & Gopon, P. (2024). The geodynamic significance of continental UHP exhumation: New constraints from the Tso Morari Complex, NW Himalaya. *Tectonics*, 43(5). https://doi.org/10.1029/2023TC007976.

Marshak, S. (2019). *Earth: Portrait of a Planet* (6th ed.). New York: W.W. Norton & Company.

Martin, C. R., Van Buer, N. J., Matchette-Downes, H., Mueller, P. A., Cruz-Uribe, A. M., van Tongeren, J. A., Hanchar, J. M., Upadhyay, R., & Jagoutz, O. (2025). The geology of the Shyok suture zone: Evidence for Cretaceous extension of the southern Eurasian margin and Eocene India–Eurasia collision. *Journal of the Geological Society*, 182(2), jgs2024-082. https://doi.org/10.1144/jgs2024-082.

McElroy, R., Cater, J., Roberts, I., Peckham, A., & Bond, M. (1990). The structure and stratigraphy of SE Zanskar, Ladakh Himalaya. *Journal of the Geological Society*, 147(6), 989–997. https://doi.org/10.1144/gsjgs.147.6.0989.

Petterson, M. G. (2010). A review of the geology and tectonics of the Kohistan island arc, north Pakistan. *Journal of the Geological Society of London, Special Publication*, 338, 287–327. https://doi.org/10.5194/essd-14-4171.

Petterson, M. G. (2019). Plutonic and volcanic rocks of Kohistan, North Pakistan: Implications for island arc genesis and evolution. In P. J. Treloar & M. P. Searle (Eds.), *The Himalayas: A Modern Synthesis*. Special Publication of the Geological Society of London, No 483, pp. 123–164. https://doi.org/10.1144/SP483.4.

Petterson, M. G. (2023). *Himalayan Thick-Skin Basement Deformation of the Ladakh Batholith, Leh-Ladakh Region, NW India*. Cham: Springer Nature (Geology) Publications.

Searle, M. P. (1983). Stratigraphy, structure and evolution of the Tibetan–Tethys zone in Zanskar and the Indus suture zone in the Ladakh Himalaya. *Transactions of the Royal Society of Edinburgh: Earth Sciences*, 73, 205–219. https://doi.org/10.1017/S0263593300009688.

Searle, M. P. (1986). Structural evolution and sequence of thrusting in the High Himalayan, Tibetan-Tethys and Indus suture zones of Zanskar and Ladakh, Western Himalaya. *Journal of Structural Geology*, 8, 923–936. https://doi.org/ 10.1016/0191-8141(86)90037-4.

Searle, M. P., Larson, K. P., & Stipska, P. (2024). The making of Mt Everest: Channel flow and low-angle normal faults. *Journal of the Geological Society*, jgs2024-118. https://doi.org/10.1144/jgs2024-118.

Treloar, P. J., & Searle, M. P. (Eds.). (2019). *Himalayan Tectonics: A Modern Synthesis*. Geological Society, London, Special Publications, 483, 215–254. https://doi.org/10.1144/SP483.7.

Indian Plate Geology *A Field Geological Laboratory Insights Approach to Deep Crustal Processes and Rock Deformation*

4

HIGHER HIMALAYA CRYSTALLINE ROCKS AND LANDSCAPES

Figure 4.1 presents a simple geological map of Zanskar, southern Ladakh. The oldest rocks (visible in the SW part of Figure 4.1) are the High Himalaya Crystallines, which form the basement geology of the NW Indian Plate. High Himalaya Crystallines around Nanga Parbat, NW of Ladakh in Pakistan, date back to around 2 billion years old (Butler, 2019), but more widely they vary in age to as young as Palaeozoic (Searle et al., 1988). High Himalaya Crystallines *protolith* lithologies (original pre-metamorphic lithologies) are dominated by siliciclastic sediments and minor carbonates, with igneous granitic and basaltic components. Whilst the deformation and metamorphism of the High Himalaya Crystallines occurred mostly during the Himalayan Orogeny, from 45–5 Ma (Searle, 1983, 1986, Searle et al., 1988), including partial melting and production of anatectic Mio-Pliocene leucogranites from 25 Ma to as young as <1Ma (Searle et al., 1988, Butler, 2019, Treloar and Searle, 2019), it is probable that the Crystallines have undergone pre-Himalayan metamorphic-thermal deformation events. Figure 4.1 presents the geographical distribution of the Zanskar Sequence (Lower and Upper, separated by the Panjal Trap Volcanics), which is aged between PreCambrian and Lower Tertiary, and forms the cover sequence of the Indian Plate in Zanskar. The Zanskar Shear Zone (Figure 4.1) separates the High Himalaya Crystallines from the Zanskar Sequence. The Zanskar Shear Zone trends NW-SE over 80–150 km of country and facilitated the rapid uplift of deep crust upwards to the Earth's surface. A steep metamorphic gradient exists across the Zanskar Shear Zone as it separates rocks of amphibolite-granulite grade within the High Himalaya Crystallines from greenschist and lower metamorphic grade Zanskar Sequence rocks (Searle et al., 1988).

The High Himalayas form some of the most magnificent landscapes of the Greater Himalaya and include almost all of the 8,000 m plus peaks. In Zanskar, the High Himalaya is characterised by jagged peaks, serrated skylines, deep valleys, and remarkably planar-linear junctions between the northern boundary of the mountain range within Zanskar. Figure 4.2 presents a view of the High Himalaya mountains, looking south from a ledge above the Padum Valley, north of Karsha. In this photograph, the sheer rise of the mountains from the 3,500 m high Padum Plain is clear, with numerous snow-capped pointed peaks jutting into the sky, providing a 'home' for small glaciers and permanent ice accumulations. A

DOI: 10.1201/9781003794868-4

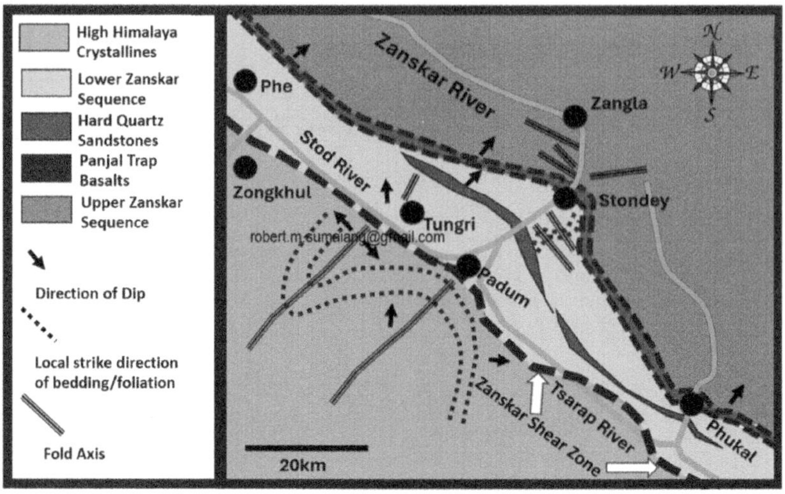

FIGURE 4.1 Geological map of the Indian plate geological units of Zanskar. *Source*: Author's map. Note: The Zanskar Shear Zone, which separates the Higher Himalaya Crystallines from the Zanskar Sequence (in turn divided into Lower and Upper Zanskar Units separated by the Panjal Trap Volcanics).

FIGURE 4.2 Looking southwards from the Padum Valley, Zanskar, across the Zanskar Shear Zone towards the spectacular High Himalaya Crystalline Peaks. *Source*: Author's photograph. Note: The glacial Hoftel Valley cuts the High Himalaya, forming a steep 'U' shaped valley.

prominent deep 'U' shaped valley (the Hoftel Valley) carved by historic glaciers, with a *misfit river* winding its way through glacial deposits, is visible at the centre of the figure.

One remarkable feature of the High Himalaya in the Zanskar region is the planar-linear form of the junction between mountains and plain, and a dipping escarpment feature which dips (tilts) at 30°–40° towards the plain from the mountain peaks (Figure 4.3). This landscape feature extends many kilometres locally from the base of Shinku-La to the western edge of the Padum Plains. This feature is the topographic expression of the Zanskar Shear Zone. Figure 4.3A presents a 'window' through the shear zone which exposes the actual contact between the Higher Himalaya Crystallines and the lowermost Zanskar Sequence. Figure 4.3B gives an appreciation of the sheer size, scope, and scale of the planar topographic escarpment, present on the northern slopes of the High Himalaya, extending for many kilometres.

The crustal-scale Zanskar Shear Zone probably had a horizontal as well as a vertical motion. Some researchers hypothesise that the Padum Plain was created by the movement of the Zanskar Shear Zone, causing a topographic low to form between the Higher Himalaya and Zanskar Ranges, which was later

FIGURE 4.3 A) displays the fault contact between the Higher Himalaya Crystalline Rocks and the Zanskar Sedimentary Rocks. B) presents the scale and extent of the remarkable linear/planar nature of the Zanskar Shear Zone, and the abrupt change from mountains to plains. *Source*: Author's photograph.

FIGURE 4.4 View looking south and east across the Padum Plain in Zanskar to the linear form of the High Himalaya Mountains, Zanskar Shear Zone, and the abrupt linear junction between mountain and plain. *Source*: Author's photographic montage.

infilled by alluvial and glacial sediments (Herren, 1987). Figure 4.4 is a composite of a number of images to produce a panoramic view of the Padum Plain from Karsha Gompa. This image illustrates populated plains with thousands of trees planted by villagers, suddenly giving way to the escarpments of the High Himalaya, cut by many conical-fan-like sedimentary deposits called *alluvial fans*: views such as this typify the dramatic landscapes of the northern margin of the Ladakhi High Himalaya.

DEEP EARTH PROCESSES AND THE FORMATION OF HIGHER HIMALAYA CRYSTALLINES

To understand the nature of the Higher Himalaya Crystallines, we have to delve into deep Earth processes. Higher Himalaya Crystallines are *metamorphic rocks* (readers can explore further details of metamorphism in Chapter 8 of Marshak, 2019). Metamorphism literally means a change (meta) of shape and form (morphism). A biological analogy is the emergence of a butterfly from a caterpillar that makes a chrysalis and miraculously changes shape from a crawling creature to a flying beauty. Figure 4.5 illustrates how we can begin with plain mud, subject it to increasing heat and pressure, and cause it to change

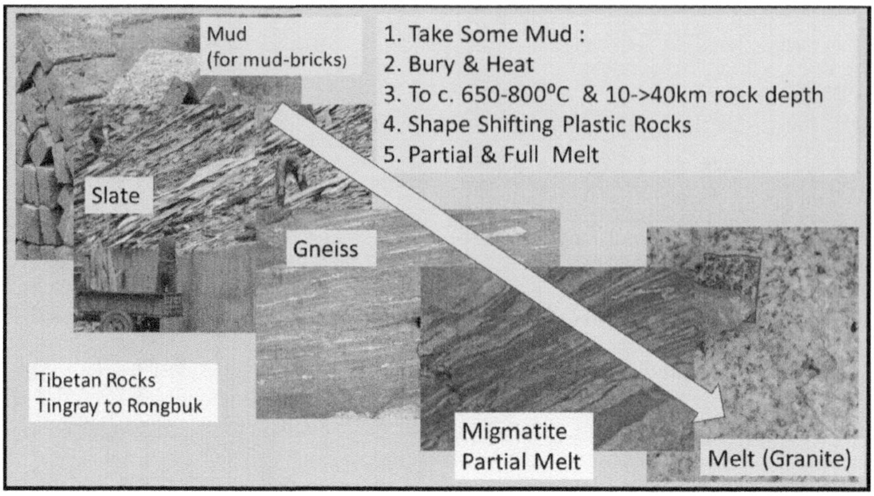

FIGURE 4.5 A metamorphic journey from mud to slate to gneiss to migmatite and granite. *Source*: Author's diagram.

into rocks such as *slate* (hard rocks that are fissile and can be split into thin layers for roofs), *gneiss* (a coarse-grained crystalline rock with a strong layered fabric called foliation), and *migmatite* (a layered rock that has experienced partial melting: up to 50% of the rock melts and generates white granitic layers that remain within the rock mass). The character of a gneiss is fundamentally different from that of a slate or a mud. The original non-metamorphosed material is termed the *protolith,* which is subject to heat and pressure, along with other variables such as fluid activity. Mineral and chemical reactions take place in the rock within the solid state (although very high temperatures cause rocks to melt). As the grade of metamorphism increases, new minerals are produced, indicative of the temperature and pressure environment in which they form. Minerals and rocks become more ductile/plastic with higher grades and can change shape.

Green minerals such as chlorite and epidote form from temperatures of <200°C to 400°C. Micas such as muscovite and biotite form at temperatures between 200°C and 650°C. Garnet forms at temperatures of 350°C to >700°C, and sillimanite at 600–800°C (Figure 4.6). Melting of rocks begins at around 650°C, depending on the ambient pressures and the presence of water and other volatiles (which lower melting temperatures). Hundred per cent melt occurs at around 1000°C. We can recognise the grade of a metamorphic rock from its mineral assemblage and classify metamorphic rocks with labels such as *greenschist, amphibolite, and granulite grades*, or low, medium, and high-grade metamorphic rocks. The longer the rocks are subjected to high temperatures and pressures, the more mature the development of metamorphic fabrics will be, and the larger the metamorphic crystals will grow. As rocks become increasingly subject to heat and pressure, their engineering properties (or *rheology*) change. At lower temperatures and pressures, metamorphic rocks are *brittle* and fracture. At higher grades, they become *ductile* and tend to flow and shear rather than fracture. Layered fabrics termed *cleavage, schistosity, and foliation* (seen mainly in rocks such as slates, schists, and gneisses, respectively) replace original protolith fabrics and completely alter the original character of the protolith.

Metamorphism is rather abstract and complex to communicate. By analogy, pottery is something more tangible. When a potter wishes to turn her/his clay creation into a longer-lasting ceramic product, s/he must metamorphose her/his clay. To do this, the clay is placed inside a furnace/oven and 'fired' at high temperature. It can take hours or even a day or so for the clay to miraculously turn into ceramic material: a pot, a vase, a tile, a brick, a jug, a plate. The fired product material is quite different from the original clay. Whilst in the furnace, the clay dries and undergoes mineral reactions: new metamorphic minerals replace the original clay minerals and provide the hardness and sustainability ceramics require for longevity and durability.

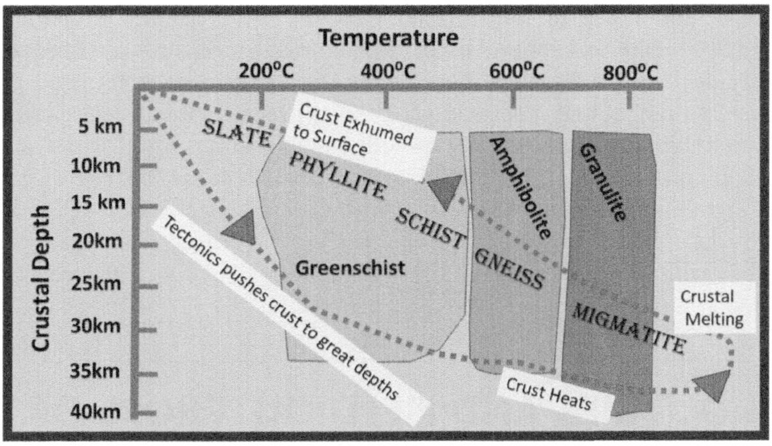

FIGURE 4.6 Pressure-temperature-depth earth environments for different grades of metamorphism. *Source*: Author's Diagram. Note: The dotted line shows the pressure-temperature-time trajectory of a possible Higher Himalaya Crystalline rock.

Figure 4.6 provides clues about what happens to the Earth's crust as it is transformed from a near-surface setting to a deep-Earth setting, and returns to the surface. Greenschist, amphibolite, and granulite metamorphic grades are defined within pressure-temperature space. Greenschist grade rocks are heated between 200°C and 500°C and form mostly at crustal depths of 5–15 km. They can form at deeper levels if the rate of change of temperature with depth (termed the *geothermal gradient)* is low, or shallower levels if the geothermal gradient is high. Greenschist grade rocks contain minerals such as chlorite and epidote, and undergo significant dewatering, which assists the formation of abundant quartz/silica veins, and develops strong planar fabrics called schistosity, producing *schists.* Amphibolite-grade rocks form from 5 km if geothermal gradients are high, but more normally form between 12 and 25–30 km. They remain hydrous, containing the mineral amphibole, which contains water, alongside anhydrous minerals such as garnet, muscovite, and biotite. Amphibolites have strong planar schist fabrics. Granulites form normally at depths of 20–40 km but can form at shallower depths dependent on geothermal gradients and tectonics. Granulites are anhydrous, tend to be coarse-grained, and form 'chunky' planar foliation fabrics, producing *gneiss.* Minerals such as feldspar, garnet, pyroxene, and sillimanite are typical of granulite assemblages. As the grade becomes very high, rocks begin to partially melt, producing migmatites composed of igneous granitic layers alongside the parts of the rock that have not melted, giving it a stripy/banded appearance. Ultimately, rocks melt completely, moving into the igneous rock space. Figure 4.6 presents a hypothetical depth (pressure)-temperature-time trajectory of a piece of crust within a High Himalaya Crystallines situation. As a general rule, every 3 km of rock is equivalent to approximately 1,000 atmospheres, or one kilobar, in terms of pressure. The Himalayas are subject to high-magnitude crustal deformation capable of rapidly pushing crust downwards to significant depths: perhaps even as deep as 90 km. Mostly, Himalayan deformation processes linked to the continental subduction/underplating of India will take significant volumes of crust to around 35 km. As it moves downwards, and becomes subject to increased heat and pressure, the original surface rock (let's say a mud rock or shale) will markedly change its character and mineralogy. Rocks are insulating materials, taking time to heat up and adopt the ambient conditions of the surrounding environment. The longer they remain within the deeper, hotter environments, the greater the metamorphic transformation. Once the crust is buried to depths of 30–40 km, it heats up and transforms, moving through the greenschist-amphibolite grades, ending up as a granulite gneiss, and ultimately experiencing partial melt, turning it into a migmatite. Part of the crust will be totally melted: this may form a large enough body of magma that separates from the residual host rock and ultimately rises in the crust, freezing as a granite intrusion. We can observe the Higher Himalaya Crystallines at the Earth's surface: hence, as Earth processes have pushed the crust to deep and hot zones within the Earth, transforming the crust into crystalline gneiss and migmatite, they have also exhumed deep crust back to the surface, again through structures such as deep shear zones and shallower reverse faults (like the Zanskar Shear Zone). Once metamorphic minerals and fabrics form at high grade, they are 'frozen' at that state unless they sit at lower grades for long periods of time. In this hypothetical situation (which reflects reality), the crust is moved quickly through lower grades, reaching the surface as granulite-grade gneiss and migmatite. The whole sequence of events described above and presented in Figure 4.6 may take several millions of years. In the case of the Higher Himalayan Crystallines, the rocks were deformed and transported via the crustal scale Main Central Thrust and related thrusts, along with the Zanskar Shear Zone, which were most active between 25 and <15 Ma.

LITHOLOGIES AND FIELD OUTCROPS OF THE HIGHER HIMALAYA CRYSTALLINES

The High Himalaya Crystallines (HHC) are composed of a wide range of lithologies. The protolith (rock type before it was metamorphosed) was mostly a mixture of mudrocks (shales), sandstones, and siltstones. The resultant schists, gneisses, and migmatites look very different. At an outcrop scale, the HHC form strong, upstanding topographic features such as the crags and cliffs of Zonghul Valley in Zanskar (Figure 4.7). The HHC form *escarpment features*: escarpments which are upstanding ridges with a steep slope (*scarp slope*) and an angled slope (*dip slope*) that reflects the geological dip of the rocks. Scarp and dip slopes meet at the *crest* of the escarpment. The *dip of a rock* is the angle a rock plane (bedding plane, schist plane, foliation plane) makes to the horizontal. Figure 4.7 displays a series of dipping rock units that dip at 30° to 50° to the right. The Zanskar Shear Zone is also exposed, which is a faulted contact between the High Himalaya Crystallines and the Zanskar (Sedimentary) Sequence. The HHC comprise thick and thin layers of schist, gneiss, and granite. Lighter coloured layers are granite sheets and quartz-rich layers.

Figure 4.8 presents typical examples of HHC Himalaya gneisses. Both pictures present light coloured (*leucocratic*) rocks: light minerals are much more abundant than dark (*melanocratic*) minerals. The lower picture presents large white feldspar crystals set within a much finer grained groundmass of quartz, feldspar, and mica: this texture is termed *porphyroblastic*. The upper photograph displays a strongly foliated gneiss: crystals have been extended in one direction and compressed in another. This striking fabric is formed under intense pressures and high temperatures. Rocks are deformed in a ductile manner and force

FIGURE 4.7 Outcrop view of High Himalaya Crystallines, showing the banded character of the gneisses and migmatites. *Source*: Author's photograph. Note: Pale-cream-white layers are granite sheets. The Zanskar Shear Zone is exposed as the contact between the High Himalaya Crystallines and the Zanskar Sequence. Zonghul Valley, Zanskar.

FIGURE 4.8 High Himalaya Crystallines Gneisses. *Source*: Author's photographs. Note: Upper photograph: highly (thinly) foliated gneiss dominated by light feldspathic foliations, with thin foliations composed of darker minerals. Lower photograph: large feldspar crystals set within a finer-grained foliated groundmass. This texture is termed porphyroblastic. The compass in upper photograph is 12cm long.

crystals to align according to the local three-dimensional stress field. Crystals which display this sort of fabric exhibit a *preferred orientation*. These rocks are high (granulite) grade metamorphic rocks: Figure 4.6 suggests formation temperature of >650°C and crustal depths of 20–30km.

Figure 4.9 develops the theme developed around Figure 4.8. Both lithologies are porphyroblastic in texture with larger feldspar crystals (termed *porphyroblasts*) displaying strong preferred orientations. The ratio of their long: short axes is c.3–5. This crystal shape has developed under a strong ductile stress field. Crystals align and form planar fabrics termed *foliations*. The porphyroblasts have 'tails' that grow from the upper right and lower left of the main crystal body. The finer-grained darker background material (termed the *groundmass*) swirls and wraps around the larger crystals. Shear stress acts in two ways: *pure shear* compresses, whilst *simple shear* exerts horizontal forces that move the upper parts of crystals to the right and lower parts to the left. These types of structures allow geologists to measure tectonic movement in 3-D space: for example, to the north or to the east. Mapping of these *shear vectors* allows geologists to model how the crust moved, whilst the rocks were being deformed and metamorphosed.

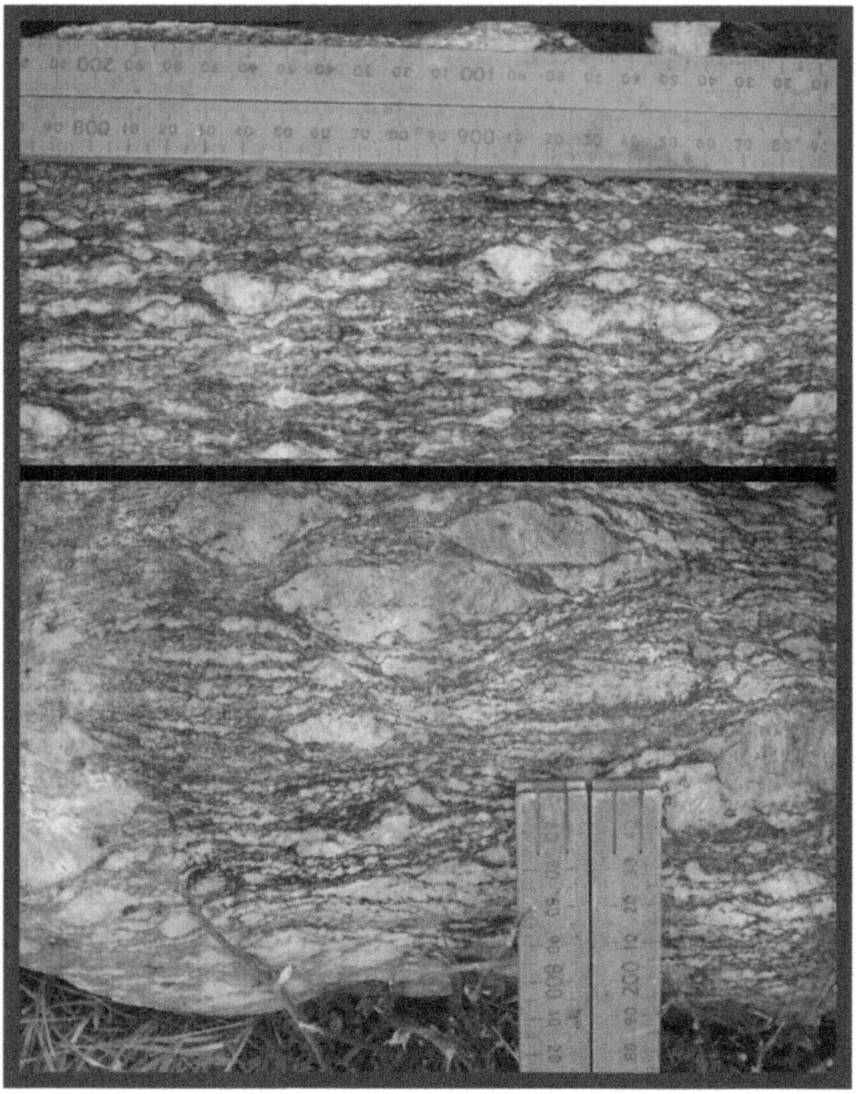

FIGURE 4.9 Strongly foliated High Himalaya Crystallines Gneisses. *Source*: Author's photographs. Note: Both photographs display strong preferred orientations, porphyroblastic textures, and 'tails' to the porphyroblasts, evidencing strong shear forces shaping rock textures.

Figure 4.10 presents lithologies of a different character to the gneisses in Figures 4.8 and 4.9. Rocks in Figure 4.10 are finer-grained and cut by thick white quartz-feldspathic veins. The rock groundmass is fine to medium-grained, and has a salt-and-pepper, black-and-white appearance. Careful examination of the background reveals a fabric that has a strong preferred orientation, which is planar to curvilinear in form. Minerals within the groundmass are predominantly quartz and mica. Mica has a plate-like form, which readily bends its shape to the local stress field, and forms the key orientated mineral within the rock fabric. Quartz is 'less compliant' to stress but ultimately stretches, compresses, and recrystallises, also forming predominant rock fabrics. This fabric is termed *schistose*, and rocks that display schistose fabrics are termed *schists*. Figure 4.6 suggests that schists form at pressures and temperatures within greenschist and amphibolite metamorphic grades. The upper photograph in Figure 4.10 presents thicker quartz layers that are an integral part of the rock mass and fabric. A careful examination of the upper boundary of the larger quartz layer reveals the presence of a black mineral layer: this is termed a *reaction rim*

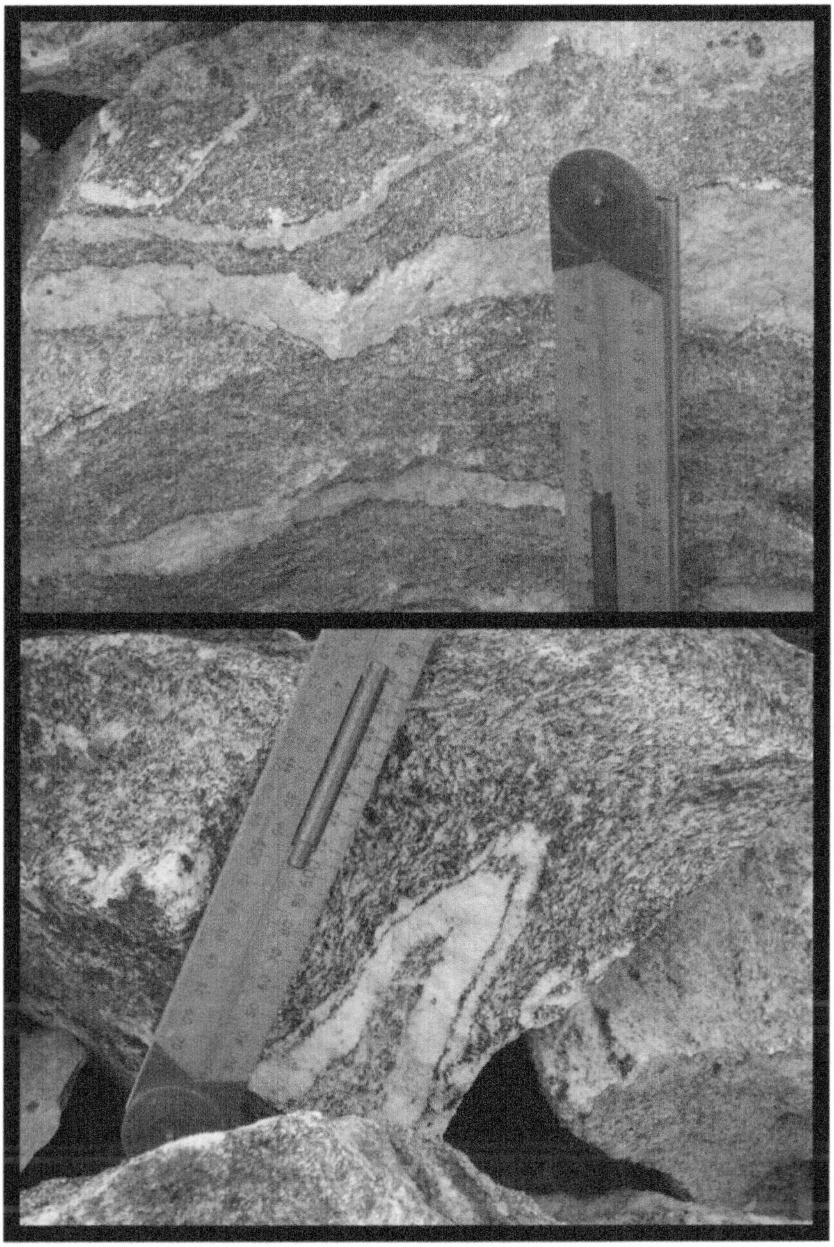

FIGURE 4.10 HHC schists. *Source*: Author's photographs. Note: These metamorphic rocks have a different character from Figures 4.8 and 4.9. Schists are dominated by phyllosilicate minerals such as micas, which form more thinly foliated rocks and well-defined foliations. Thicker sheets of quartz-feldspar cut the schistose fabrics: the lower photograph presents a tight (isoclinal) fold.

or boundary, revealing chemical and mineralogical reactions that take place under metamorphism: the boundary between the quartz layer and the groundmass was a fertile zone of various chemical reactants that developed new minerals, in this case mica and amphibole. Metamorphic reactions occur in the solid state, although the presence of volatiles expedites reactions. The lower photograph in Figure 4.10 presents a strongly folded quartz layer within the rock mass. Quartz at surface temperatures and pressures is highly brittle, and could not bend and flow as Figure 4.10 indicates. This quartz layer has been halved in its length, doubled in its thickness, and shape-shifted into a 'U' shaped fold structure (termed *isoclinal*). Structures like this fold provide valuable insights into the conditions within which the metamorphic rocks formed: at temperatures of around 500ºC, and depths of 15–20 km (pressures of 5–7 kilobars, equivalent to 5,000–7,000 atmospheres). Ductile rock materials were then malleably deformed and recrystallised. Structures such as the fold in Figure 4.10 help geologists determine crustal movements at the time of metamorphism, which in turn leads to the modelling of tectonic plate movements through time.

Figure 4.11 provides more clues about the original protolithic materials of the High Himalaya Crystallines. The lower photograph displays a c. 4 cm thick white layer, dominated by medium-grained quartz crystals and thin dark micaceous (formerly muddy) layers. This layer is a meta-quartz rich sandstone with minor mudstone layers. The upper photograph displays a more deformed and recrystallised sequence of meta quartz sandstones with interbedded mudstones: now quartz-mica schists. The sparkly white mineral in the upper photograph is muscovite mica. The red tint to the quartz layer indicates mobility of iron involved within the metamorphic recrystallisation of quartz.

Figure 4.12 presents an upper quartz-mica schist with tourmaline mineralisation, and a lower schist with strong 'S' shaped fabrics. Tourmaline forms black prismatic columnar-like minerals with angular and tabular cross sections. Tourmaline indicates the activity of hot aqueous solutions in liquid or gaseous form, along with other volatiles, particularly boron. Tourmaline has a complex composition including elements such as calcium, sodium, potassium, lithium, magnesium, iron, and boron. In the Figure 4.12 example, tourmalines are spatially associated with quartz layers and the boundary between the quartz layer and the mica-quartz finer-grained groundmass. This is the site of fluid movement, which transports reactants and heat energy in a volatile form. The lower photograph in Figure 4.12 presents a simple-shear-deformed rock which has formed 'S' shape fabrics in response to the shear. The larger crystals and crystal aggregates/quartz pods have a strong 'S-shaped' wavy geometry with tapering tails along the top right and lower left of the crystal/pod (see the quartz pod next to the pencil in the photograph). The finer-grained mica-quartz schistose layer wraps around the larger crystals and masses of quartz. Careful observation of the finer-grained groundmass reveals a swirling fabric, with many examples of local fabrics that lean to the right-hand side of the photograph. An analogue of simple shear deformation can be used to explain how this works. If we undergo a 'thought experiment' and imagine a pack of playing cards: the bottom of the cards in our left hand supports the bottom of the deck of cards, whilst our right hand rests on the surface of the deck of cards. We then move our left hand to the left and our right hand to the right. In response to this simple shear, the cards move: the lower part of the deck to the left and the upper part to

FIGURE 4.11 Upper quartz-mica schist and lower meta-quartzite. *Source*: Author's photographs. Clues to protolithic materials of the HHC. 7cm of pencil exposed in these images for scale.

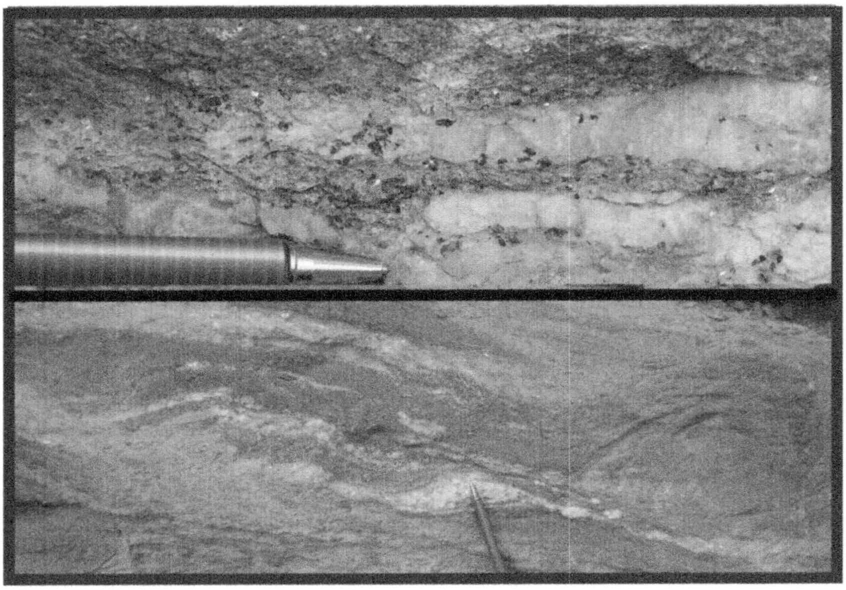

FIGURE 4.12 Upper quartz-mica schist with tourmaline (black) crystals and lower schist displaying S-shaped 'wavy' sheared foliation structures. *Source*: Author's photograph.7cm of a pencil is exposed in both images.

FIGURE 4.13 Staurolite-muscovite schist. *Source*: Author's photograph. Note: Large brown staurolite porphyroblasts with rectangular prismatic geometries are set within a finer-grained, muscovite-dominated groundmass. 2cm of a pencil included in the image for scale.

the right, forming a parallelogram shape. Similarly, rock masses within a ductile environment respond to simple shear within deep crustal ductile environments. In this case, we would describe the shear as 'top to the right'. Geologists can measure these structures in three dimensions at outcrop scale and then mark these measurements geographically across the study area. A pattern of movement will emerge from the mapping of these structures, which can then be used to model three-dimensional crustal movements and deformation in space and time.

Figure 4.13 presents an excellent example of a metamorphic rock with an *indicator mineral*: in this case, staurolite. An indicator mineral provides strong evidence for the metamorphic grade of the rock in which it forms. Staurolite is an iron alumino-silicate often presenting prismatic crystals, as shown in Figure 4.13. Staurolite can form over a range of pressures and temperatures, but normally indicates high-grade metamorphism. Temperatures of >500°C to >650°C and pressures equivalent to 20- > 30 km crustal depths have been estimated in a range of laboratories (Marshak, 2019). Depending on pressure-temperature-fluid activity and available reactant conditions, staurolite can form in association with other minerals such as garnet, chlorite, muscovite, biotite, and sillimanite. In Figure 4.13, staurolite presents well-formed (euhedral) brown crystals, with a strong rectangular prismatic habit. Individual crystals are up to 20 mm long. The large staurolite porphyroblast crystals set within a finer-grained muscovite-dominated groundmass display a well-developed porphyroblastic texture.

Figure 4.14 contains abundant red garnets, sometimes found within the gneiss groundmass and also surrounded by white felsic material. Black tourmaline is also present. Like staurolite, garnet indicates pressure-temperature formation conditions, in this case from >500°C to >700°C and at crustal depths between 15–20 and >30–35 km. This gneiss displays a porphyroblastic texture with large white porphyroblastic feldspar crystals set in a finer-grained matrix. Porphyroblasts are lozenge-shaped due to simple shear and grow linear 'tails' along the upper right and lower left extremities of the lozenge crystals. These textures help geologists determine the local sense of simple shear, here from top to the right. A contrast in grain size between the lower and upper part of the photograph indicates initial contrasting grain sizes of the original protolith, with alternate fine- and coarse-grained muddy siltstones and sandstones.

Figure 4.15 presents close-ups of the silica-rich, white-pale felsic components of High Himalaya Crystallines. The uppermost photograph presents a red garnet-rich quartz vein, with the middle and lower photographs displaying black tourmaline crystals within quartz layers. The lowermost photograph shows lenticles of quartz that emanate from the main vein and bend to the right in response to simple shear, indicating top to the right local shear sense. Quartz alone can neither produce garnet nor tourmaline under metamorphic conditions, as quartz is a simple chemical system composed of SiO_2. Garnet and tourmaline require many other elements such as iron, magnesium, and aluminium. The quartz may react with the muddier parts of the rock mass. The role of fluids and volatiles is often linked with quartz veins, as SiO_2 is highly mobile with hot water (hydrothermal) solutions. Sometimes gold and copper may also form as an ore body within such environments.

Figure 4.16 presents spectacular examples of events that happened during the final stages of the High Himalaya Crystallines metamorphism. The striking appearance of migmatite mountains is awe-inspiring to all, particularly to those who have some geological knowledge. The stripy black and white rocks, composed of myriads of lines, layers, cross-cutting veins, and large masses of white granite, dominate local outcrops. Black and dark-grey coloured gneisses and schists form the host rock, contrasting markedly

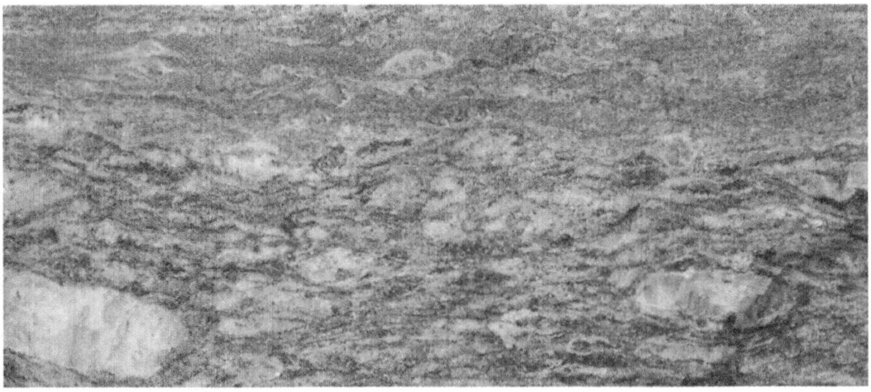

FIGURE 4.14 Sheared garnet-feldspar-mica gneiss. *Source*: Author's photographs. Note: The rock displays a porphyroblastic texture with porphyroblasts (large crystals) of feldspar and quartz. Porphyroblasts display 'fish-textures' with linear tails extending from the upper right and lower left porphyroblast extremities, indicating top to the right shear sense. Red round garnet crystals are present in the groundmass and within reaction rims around some porphyroblasts. Large white feldspar crystal, lower left is 2cm long.

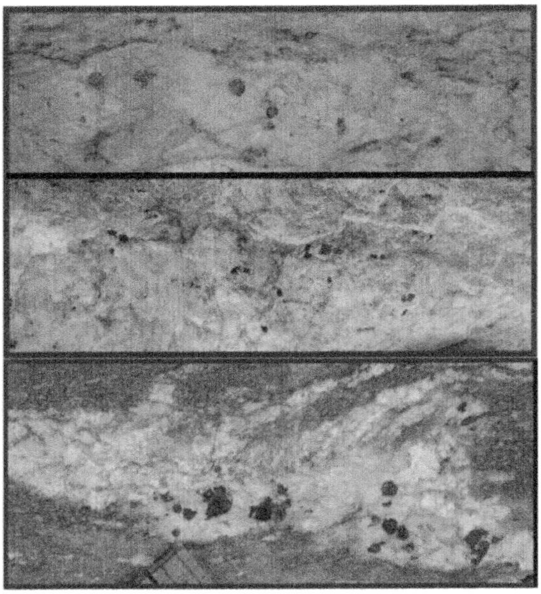

FIGURE 4.15 Upper garnet-rich felsic gneiss with tourmaline crystals abundantly present in the middle and lower photographs. *Source*: Author's photographs. Red garnets in upper photograph measure 5-10mm. Tourmalines crystals in middle photograph measure 4-8mm. 2cm of a ruler is visible in the lower photograph.

FIGURE 4.16 Spectacular migmatite mountains from the Gombo Rangjon area of easternmost Zanskar. *Source*: Author's photographs. Note: Migmatites, literally 'mixed rocks', comprise residual dark coloured granulite grade gneisses and abundant white leucogranite sheets, which intrude their host rock (melted to produce leucogranites) in dense criss-crossing (lit-par-lit) networks.

with the white leucogranite layers. These mountains mark the latter stages of HHC-Himalayan metamorphism, moving from purely metamorphic processes where reactions take place in the solid state to a part-metamorphic and part-igneous domain: of liquid rock, magmas, and the molten state. Migmatites are mixed rocks: part-metamorphic and part-igneous. By the time migmatites form, deep crustal conditions have reached a high metamorphic culmination: the point where the *solidus* is exceeded. The solidus is the term for the upper boundary where rocks are completely solid: once the solidus is exceeded, the rock becomes partly or wholly liquid. As granulite-grade metamorphic rocks melt, they do not melt en masse. Rock constituents that have the lowest melting points melt first. This produces a magma rich in SiO_2 or silica. These siliceous magmas move from the point at which they are produced and amalgamate with other small volumes of melt to create a larger magma body. These magma bodies then move through the rock mass: along voids, cracks, and vortices, and deform the remaining solid rock mass as it migrates.

As the whole temperature of the system is increased, the highly ductile nature of solid components can readily deform. Within migmatites, magmas do not move far and freeze within the rock mass in which they were produced, moving metres to hundreds of metres. The solid rock that has not melted is termed the residual material or *restite*.

Figure 4.17 displays the rock *leucogranite*, which forms from the lowest temperature melting of crustal rocks. They are termed leucogranite, referring to the relative paucity of dark minerals and are mostly composed of feldspar and quartz. The glassy (vitreous) crystals in Figure 4.17 are quartz, and the milky white crystals are feldspar. To the left of the pencil point are red garnets. Tourmaline and muscovite also form a small component of the Gombo Rangjon (easternmost part of Zanskar) leucogranites and migmatites. The whole Himalayan region contains crustal melt leucogranites produced at the height of movement of the Main Central Thrust, mostly formed between 25 and 5 Ma (Searle et al., 1988).

Figure 4.18 presents a typical outcrop scale example of the High Himalaya Crystallines. At this scale, the heterogeneity of the gneisses and schists is visible with dark- and light-coloured components. The darker units are dominated by quartz-mica schists and gneisses. Lighter coloured layers comprise granites and quartz-feldspar gneisses. The dark schists vary dramatically with respect to the thickness of their layers, as they undergo plastic deformation and pure (and simple) shear, which stretches and squeezes layers in a highly ductile manner. The dark schists are locally sheared/thrusted over the lower layer close to the white arrow. Note how the unit underneath the shear structure has been folded and fractured. Later, brittle fault structures break up the outcrop.

Figure 4.19 displays an outcrop of a highly sheared leucogranite sheet. White leucogranite has been deformed by pure and simple shear, which has stretched and squashed a formerly contiguous granitic sheet. The granite now comprises discrete lozenges or *boudins*. This process is termed *boudinage* after the French word for sausage, and allows geologists to reconstruct the ancient stress regime with maximum compression orthogonal to the boudins, and maximum extension parallel to the boudins.

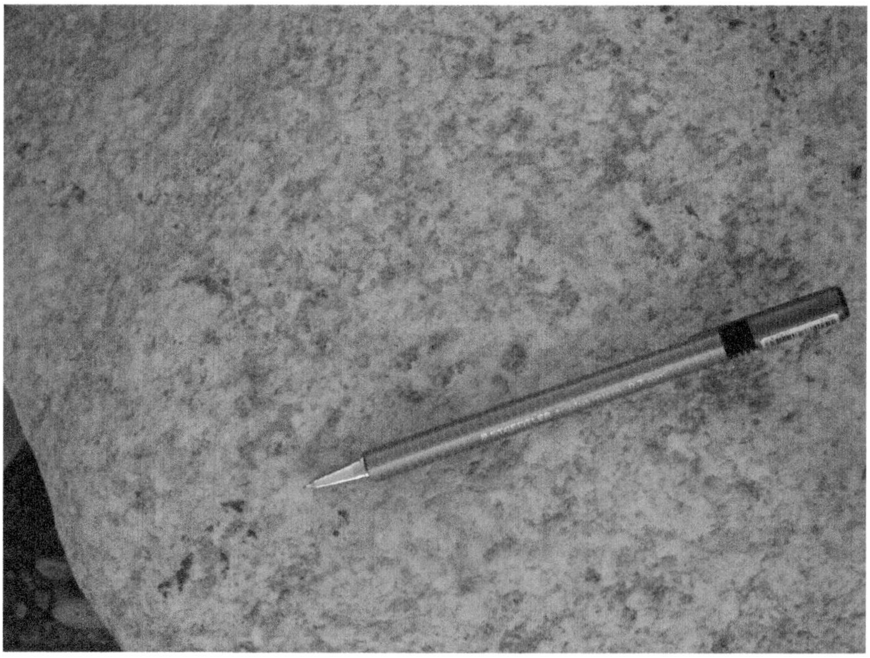

FIGURE 4.17 Leucogranite formed from crustal melting. *Source*: Author's photograph. Note: The dominance of light-coloured felsic minerals with only a very small component of dark melanocratic ferromagnesian minerals. Pencil is 15cm long.

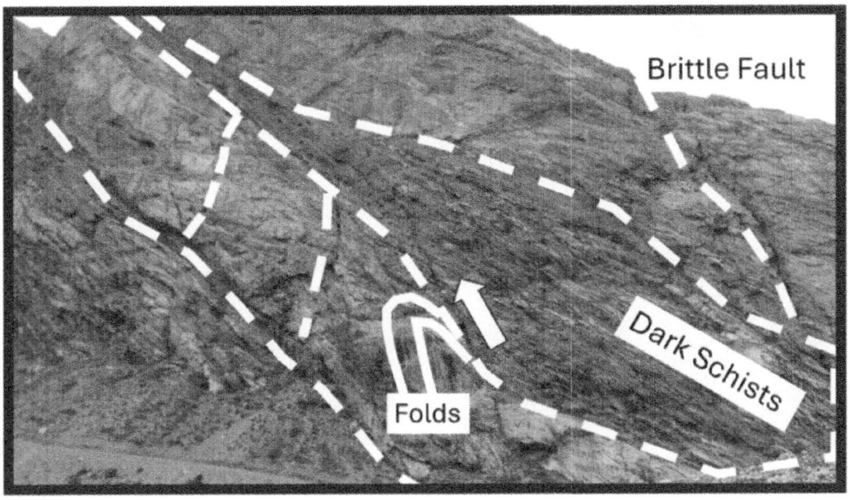

FIGURE 4.18 Mountain-scale outcrop of Higher Himalaya Crystallines in Zonghul Valley, Zanskar. *Source*: Author's photograph. Note: The alternating highly squeezed darker schists, which vary significantly in thickness, and the more rigid but fractured and folded quartz-feldspathic lighter layers.

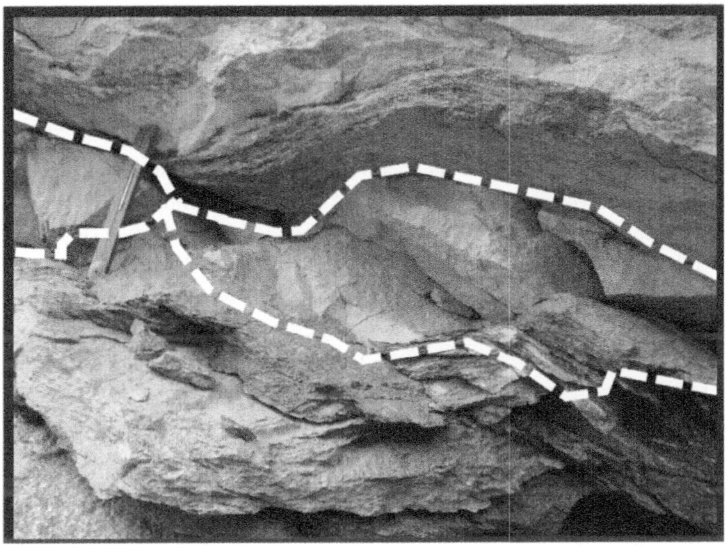

FIGURE 4.19 Sheared and deformed lozenges (boudins) of leucogranite within sheared schists. *Source*: Author's photograph. 25cm length of a metre ruler is present within the image.

Figure 4.20 (left hand side) presents a brittle *normal fault*, with the rock mass to the right moving downwards relative to the rock mass to the left (arrow indicates relative motion). Note the extensive fracturing (white dashed lines are parallel to fault and fracture structures). The photograph on the right-hand side is an example of *slickenside structures*. When faults move, they create grooves and ridges parallel to the last movement of the fault, termed *slickensides*. Geologists measure slickensides to provide information about fault motions in 3-D space.

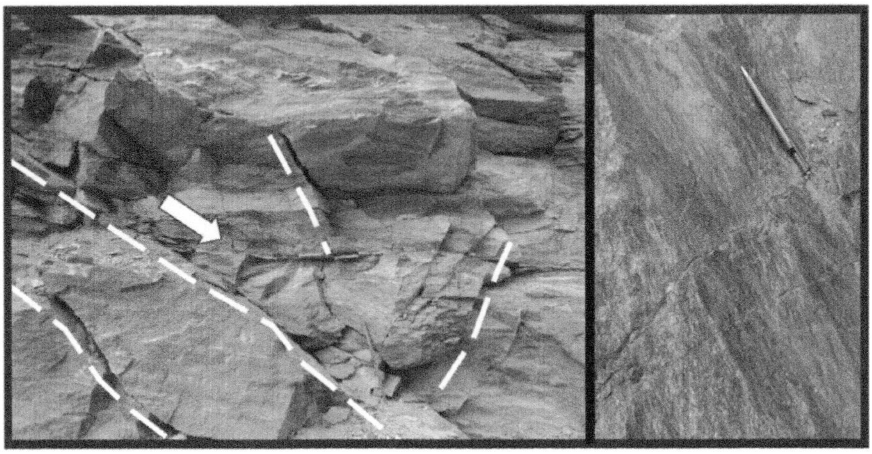

FIGURE 4.20 Left hand side: brittle faulting and fractures, with the right side moving downwards relative to the left side across a normal fault. White dashed lines placed next to, but slightly apart from fractures within the rock mass. Right Hand Side: linear structures parallel to the pencil indicate the direction of fault motion. The bulbous to tapering character of the structure is a slickenside. *Source*: Author's photographs. Left hand side photograph contains a metre ruler. Pencil in right hand side image is 15cm long.

HIGH HIMALAYA CRYSTALLINE LANDSCAPES

Gombo Rangjon (Figure 4.21) is a magnificent example of High Himalaya Crystalline Mountain scenery. The expanse of pure white leucogranite, speckled with dark schist and gneiss that refused to melt, along with sheer cliffs on all sides and its jagged peaks, combines to instil a sense of wonder. Gombo Rangjon stands alone, isolated from surrounding ridges and mountains, adding to its grandeur.

The highest parts of mountain landscapes within the High Himalaya Crystallines are dominated by the erosion and depositional environments of ice and glaciers. Figure 4.22 presents a mist-shrouded high mountain landscape devoid of vegetation and characterised by high ice-sculpted peaks. Bowl-like depressions termed *corries* form the depositional grounds of ice that build up into glaciers, should conditions allow. The peaks are surrounded by high mounds and ridges of boulders, sands, and gravels, termed *glacial moraine*, produced by melting ice, which yields the eroded material it produced. Lakes are common as the large volumes of glacial moraines dam streams and rivers, creating lakes.

The highest mountains attract the harshest weathering processes, which cut sharp edges to the peaks. As water freezes, it expands, and as it melts, it thaws (Figure 4.22). This regular expansion and contraction prises away pieces of the mountain over millions of years. Ice sculpts and scrapes moulding sharp edges to mountain ridges termed *arêtes*. Geological structure shapes mountains: in this case, the well-developed foliations of the High Himalaya Crystallines produce escarpment features with highly defined scarp and dip slopes. Life in any form struggles to survive in these extreme environments: hence the absence of greenery and the predominance of Earth colours alongside ice and snow.

Mountains are perhaps best appreciated when there is perspective provided by the lower valley setting, looking up to the peaks. Figure 4.24 presents a view looking south from the Padum Valley in Zanskar towards the High Himalaya. Lower land is dominated by plains and a braided river system, comprising a myriad of anastomosing river channels, rather than one main channel. At the heart of the picture is a deeply dissected 'U' shaped valley carved by a glacier, with steep valley sides and a broad, flat valley floor. Valleys such as these provide natural 'roads' into the heart of the mountains. Before modern vehicle

FIGURE 4.21 Gambo Rangjon: a leucogranite-dominated, isolated, pyramidal peak. *Source*: Author's photograph.

roads were constructed, people in Zanskar relied on trails that took advantage of natural pathways through the mountain fortress terrain.

River valleys within mountain landscapes are high-energy hydrological systems, rejuvenated by tectonics. As mountains are pushed up tectonically, rivers assume 'superpowers', given extra energy and strength: termed '*rejuvenation*'. Rejuvenated rivers cut deeply and quickly, creating steep 'V' shaped valleys and *interlocking spurs*. Rivers that run through particularly aggressively uplifting mountains cut down quickly, forming steep river gorges (Figure 4.25). Rivers wind their way through mountains, seeking the greatest weaknesses they can take advantage of. As a result, valley floors are narrow, sometimes fully occupied by the active river channel, forming a zig-zag mountain ridge character in plan view and creating *interlocking spurs*, with ridges alternately protruding into the river valley floor.

FIGURE 4.22 Glacial Uplands of the highest mountain terrains in Zanskar High Himalaya Crystalline Country, Shinku- La. *Source*: Author's photo-montage.

FIGURE 4.23 Upland High Himalaya looking south from Tungri, Zanskar. *Source*: Author's photograph.

FIGURE 4.24 Hoftel Glacial Valley. Zanskar. *Source*: Author's photograph. Note: The steep valley sides and wide flat valley floor forming a 'U' shaped cross-section, and the large braided river in the foreground.

FIGURE 4.25 The Tsarap River in Zanskar cuts a sharp gorge between the High Himalaya Crystallines to the right and the Zanskar Sequence to the left.

FIGURE 4.26 Well-defined escarpment country of the High Himalaya Crystallines meets bedded rocks of the Zanskar Sedimentary Sequence. *Source*: Author's photomontage. Note: The Zanskar Shear Zone downthrows the Zanskar Sequence.

The topography-forming character of adjacent differing lithologies is well-exemplified in Figure 4.26, Tsarap Valley, Zanskar. On the right-hand side, the topography is dominated by escarpments and dip and scarp slopes, reflecting the strongly foliated character of the Higher Himalayan Crystallines. On the left-hand side, the Zanskar Sequence rocks produce quite different landscapes, with more diffuse, less sharp features. Faulted contacts between the major rock groups, particularly the Zanskar Shear Zone, are well-exposed: the Zanskar Sequence is faulted downwards relative to the High Himalaya Crystallines.

LITHOLOGIES, LANDSCAPES, AND FIELD GEOLOGY OF THE NEOPROTEROZOIC-PERMIAN LOWER ZANSKAR SEQUENCE

With reference to the geological map presented in Figure 4.1. High Himalaya Crystallines crop out within the southwestern part of the map. The Lower Zanskar Sequence crops out northeast of the Zanskar Shear Zone. Within the Lower Zanskar Sequence are a number of thick, hard quartz sandstone units that weather a strong red to maroon-ochre colour. A prominent marker basaltic unit (Panjal Traps) separates the Lower and Upper Zanskar Sequences. This unit was erupted as part of a large-scale continental flood basalt event: a much larger outcrop of these volcanic rocks is present in Kashmir, west of Zanskar. The Upper Zanskar Sequence crops out northeast of the Panjal Traps basalts. Local rock dips are shown in Figure 4.1 by arrows: the regional dip is mostly to the north, but the prevalence of folding within the Zanskar Unit modifies dips locally.

The Precambrian–Permian Lower Zanskar Sequence comprises a sequence of mudstones, sandstones, quartzites, conglomerates, carbonates, and limestones. Sedimentary *facies* (a term that links sedimentary rocks and sediments to their depositional environment) for the Lower Zanskar Sequence have been interpreted as low-energy, deep-marine (mudstones); shallow-marine, carbonate shelf (carbonate-siltstones and limestones); high-energy shoreface-ocean interaction zone (quartzites, conglomerates); and deltaic-fluvial (siltstones and lithic sandstones) (Nanda and Singh, 1976; Garzanti et al., 1986a; Gaetani et al., 1986; Myrow et al., 2023). India was part of the Rodinia continent during the later Proterozoic times. Following the breakup of Rodinia, the Indian plate became part of eastern Gondwana from the Cambrian to the Cretaceous. Lower Zanskar Sequence sedimentary rocks record depositional environmental facies within a stable, tectonically quiet continent during Palaeozoic times (Bhargava, 2015). This book does not focus on stratigraphy, nor on a detailed literature study of stratigraphy, but rather on using rock outcrops as a field geological laboratory to illustrate geological processes within a mountain orogenic belt, including a description and overview of stratigraphy and lithology. Some stratigraphic context helps the reader. At a most fundamental level, the Zanskar Sequence is divided into two main units: the Lower and the Upper Sequence, separated by the Permian Panjal Trap Basalts; this book will work at this level. Most of the photographs and associated narratives within the Lower Zanskar Sequence are from the Haimanta Group (NeoProterozoic to Cambrian), and its constituent Formations, such as the Kunzam La Formation (Early to Middle Cambrian), Phe Formation (Neoproterozoic–Lower Cambrian), Karsha Formation (Middle Cambrian), and also the younger Muth Quartzite Formation (Devonian). For details, readers are directed to Nanda and Singh, 1976; Garzanti et al., 1986a, 1986b, 1989, 1996; Garzanti, 2024; Gaetani et al., 1986, 1991, Parcha and Singh, 2023, Thakkar et al., 2023, and Myrow et al., 2023.

Mudstones are the most common sedimentary rock in the geological column, comprising around 65-70% of all sediments and sedimentary rocks on Earth. Sediments are derived ultimately from the *weathering* of igneous (and metamorphic rocks). As rocks reach the Earth's surface, they are exposed to the elements of weather and climate. Rocks break down through a process called *weathering,* which manifests itself as physical, chemical, and biological processes that change the character and composition of fresh rocks. Physical weathering breaks down rocks into smaller and smaller particles through the action of ice, glaciers, rivers, rain, wind, and oceans. Cold and dry climates, such as the polar, mountain, and desert regions, are dominated by physical weathering (Marshak, 2019). Chemical weathering is particularly effective in warm, humid, wet climates such as the tropics. Chemical processes such as oxidation, hydration, dissolution, and carbonation chemically alter minerals, particularly within the surface layers of rocks (Marshak, 2019). Weathered rock surfaces look significantly different from fresh interiors, dominated by hydrated and oxidised minerals, and often taking on a grey, ochre, red, or brown colour. The thickness of the weathered zone varies: in the tropics it can extend several to tens of metres below the surface rock. Most minerals decay through chemical weathering, with clays being the most abundant product; hence, the abundance of mudrocks (dominated by clays). Biological weathering involves the action of biology through, for example, the breaking up of rocks by tree roots, lichens 'eating' surface layers of rocks, and microscopic organisms working in tandem with physical and chemical weathering, enhancing biochemical reactions. Quartz is the most resistant mineral to weathering, accumulating in deposits derived from river, coastal, oceanic, and wind processes. Muds and silts accumulate in quiet geological surface environments: for example, in lakes, river floodplains, and deep oceans/marine basins. As sediments accumulate, they form sedimentary basins that subside with the increasing weight of sediment. Sediment turns into sedimentary rock via a process called *lithification*. Lithification occurs through burial pressures and temperatures of around 50–180°C (Marshak, 2019). The movement of warm fluids within sediments precipitates cements that 'glue' sedimentary particles together. Sedimentary rocks are often 'bedded', made up of layers which represent depositional events of a tangible thickness of sediment, followed by a time gap between sedimentary layers. Limestones, formed from biochemical processes, extract calcium carbonate from oceans and seas through biochemical precipitation. This can occur primarily abiogenically, or largely through biological processes, for example, within coral reefs. Limestones form within shallow surface marine waters or within deep oceans, so long as the depths are above the *carbonate compensation zone*, around 3.5–5000 m below sea level, below which calcium carbonate dissolves (Marshak, 2019). Fossils are a constituent of many sedimentary rocks, particularly within mudstones and limestones. Fossils provide the data for evolutionary studies and evaluation of the age and palaeoecology of the sedimentary rocks at the time they formed. Readers are directed to Marshak, 2019 chapter 7 and Interlude B for more detailed accounts of sedimentary rocks and processes.

Figure 4.27 presents an outcrop overview (left-hand side) and close-up (right-hand side) of typical mudrocks within the Lower Zanskar Sequence. They are siliceous mudstones, siltstones, and fine-grained sandstones, dark grey in colour, thinly bedded, and preserve structures which indicate deep marine environmental depositional origins. Some outcrops can be *fissile*, meaning that the rocks can split smoothly along planes. Much of the Zanskar (Sedimentary) Sequence is metamorphosed, mostly between greenschist and sub-greenschist metamorphic grades. Mudrocks can become *cleaved*: cleavage planes are planes of high fissility allowing mudrocks to turn into slates, which can be quarried to produce roofing and floor materials. Cleavage forms through folding processes, which encourage the growth of low-grade metamorphic minerals that align to stress fields that produce folds. Fossils are present in some of the Zanskar mudrocks, including brachiopods and trilobites of Cambrian age (Nanda and Singh, 1976; Garzanti et al., 1986a, 1986b; Gaetani et al., 1986; Myrow et al., 2023).

Figure 4.28 presents a photograph of mudrocks that experienced extensive shear stress, with rock cleavage planar structures developing through the application of shear stress. Note the parting planes (cleavage) next to the ruler in Figure 2.34, which display a chevron morphology. The meeting point of the chevron structure is a ductile fault/shear, along which the upper rocks have moved to the right relative to the lower rocks. The abundance of clay and micaceous minerals, which have a sheet-like atomic structure,

FIGURE 4.27 Mudrocks from the Lower Zanskar Sequence exposed close to Tungri. *Source*: Author's photographs. Foliage within the left hand side image is c. 20cm long. Ruler within the right hand image is 25cm long.

FIGURE 4.28 Sheared and cleaved mudrocks, Karsha, Zanskar. *Source*: Author's photograph.

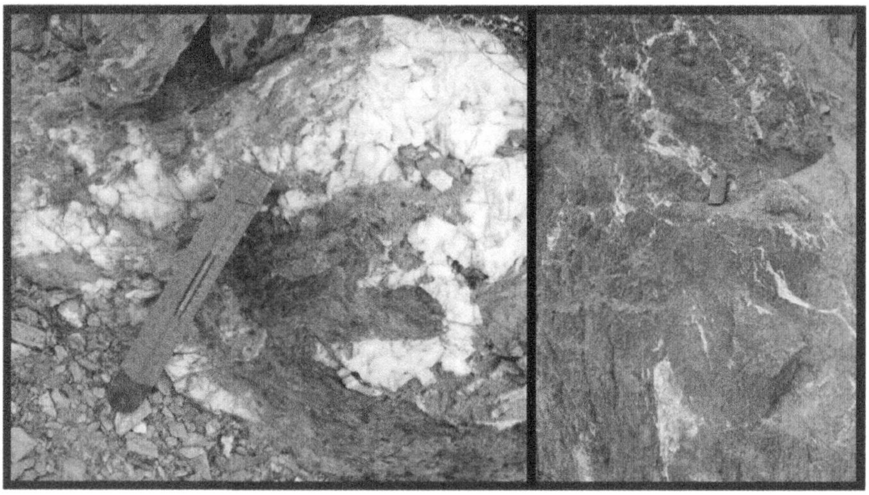

FIGURE 4.29 Abundant quartz veins within mudrocks exposed in Tungri and Karsha, Zanskar. *Source*: Author's photographs. The ruler within the left image is 25cm long, the mobile phone within the right hand image is 15cm long.

makes mudrocks prone to ductile flow under stress. Sheet silicates can 'lubricate' faults and shears, helping shears to propagate through the rock mass.

As mudrocks experience higher metamorphic temperatures, for example, within the greenschist metamorphic grade, they 'sweat'. Figure 4.29 exhibits abundant (white) quartz veins that result from the 'sweating' of mudrocks. As temperatures rise, minerals dehydrate, and the resultant hydrothermal fluids are rich in silica (in solution). As fluids move through the crust, they reach environments less conducive to retaining silica in solution, hence quartz veins precipitate within the meta-mudrocks. This process can be violent, as the hydrothermal fluids precipitating quartz are under pressure, and can literally tear the rock mass apart. This *brecciation* of the mudrocks forms angular particles (or *clasts*), which are visible in Figure 4.29. In the left-hand side of Figure 4.29, angular particles are present within the large milky white quartz vein left of the ruler, and on the right-hand side photograph, angular particles are visible close to the mobile phone. Quartz veins precipitate within hollows/cavities, and cracks/planes of opening, which tear in a brittle manner. It is within environments such as quartz-rich mudstones that gold, copper, and other metals can be co-precipitated as sulphide minerals such as chalcopyrite. This metal-rich environment is termed '*orogenic gold*' reflecting the formation of gold within mountain-building geological terranes (Marshak, 2019).

A series of hard, resistant, red sandstones forms marker beds within the Lower Zanskar Sequence. These units are quartz-rich and gritty, forming very thickly bedded outcrops (Figure 4.1). Figure 4.30 illustrates a close-up of quartz-rich gritty sandstones. On the left-hand side of Figure 4.30, the photograph presents a unit which oxidises a strong red-ochre colour caused by the oxidation of iron within the sandstone. Fresh exposures are dark light grey in colour. Abundant streaky quartz veins strengthen the rock. The right-hand side photograph of Figure 4.30 presents the coarse-grained texture of the sandstone with dark micas. A speckled appearance provides insight into the packing and coarse-grained nature of the sandstone. The lower left part of the photograph depicts the stark contrast between haematite-mineralised red weathered rock and the grey fresh character of the same material.

Figure 4.31 exhibits the bedded character of sandstones within the Lower Zanskar Sequence. The left-hand side image gives details of the alternating character of some outcrops. Thicker bedded red sandstones contain abundant thinly bedded siltstones and mudstones. The right-hand side photograph of Figure 4.31 presents medium-thick-bedded red sandstones, forming typical strong vertical cliff outcrops. Careful observation reveals that some bedding is at an angle to the general horizontal bedding attitude (e.g., next to the pencil): this sedimentary structure is termed *cross bedding* or *current bedding* and indicates that sediment was laid down under the influence of water currents, probably coastal or shallow marine currents.

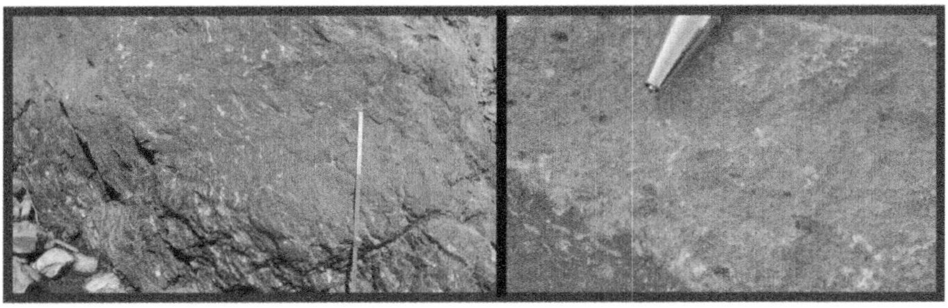

FIGURE 4.30 Red-grey coarse gritty sandstones from Karsha and Phuktal. *Source*: Author's photographs. A metre role is visible wihin the left hand side image. 2cm of a pencil is exposed within the right hand side image.

FIGURE 4.31 Alternating thickly bedded sandstones and thinly bedded siltstones (left-hand side) and thickly bedded red sandstones (right-hand side), Phuktal. *Source*: Author's photographs. Pencil visible in both images is 15cm long.

Figure 4.32 is a composite photomosaic from a series of photographs along the deep river valley towards Phuktal. The Tsarap River Valley locally trends north-northeast: the left-hand side of the figure is NNE and the right-hand side SSW. The figure includes a lower geological interpretation. The upper photomosaic presents a complete range of stratigraphy from the lower part of the Lower Zanskar Sequence, dominated by mudstones with thin sandstones, prominent red sandstones in the upper part of the Lower Zanskar Sequence, a dark-grey basalt (Panjal Trap) unit, and the Upper Zanskar Sequence, cropping out at the highest parts of the mountains. The geological interpretation provides insights helping to unravel the panorama. Two faults break up the panorama. On the left-hand side, a very thick red sandstone, amongst other strata, has a gently northward dipping disposition. Stratigraphy on the right-hand side dips much more steeply to the north. Thick red sandstones form a spectacular vertical cliff: note how the thickness of the sandstones changes dramatically across faults. Even thinly-bedded red sandstones form strong topographic features. The Panjal Trap basalt forms prominent grey-black marker horizons that can be tracked across the country. These basalts mark the boundary between the Lower and Upper Zanskar Sequences. The Upper Zanskar Sequence tends to be lighter coloured and more thinly bedded, producing terraced scarps within the highest mountain topography in Figure 4.32.

Figure 4.33 presents an example of a three-dimensional view through *chevron-folded* siltstones, within the lower part of the Zanskar Sequence. This style of folding (chevron style, similar shape to chevrons

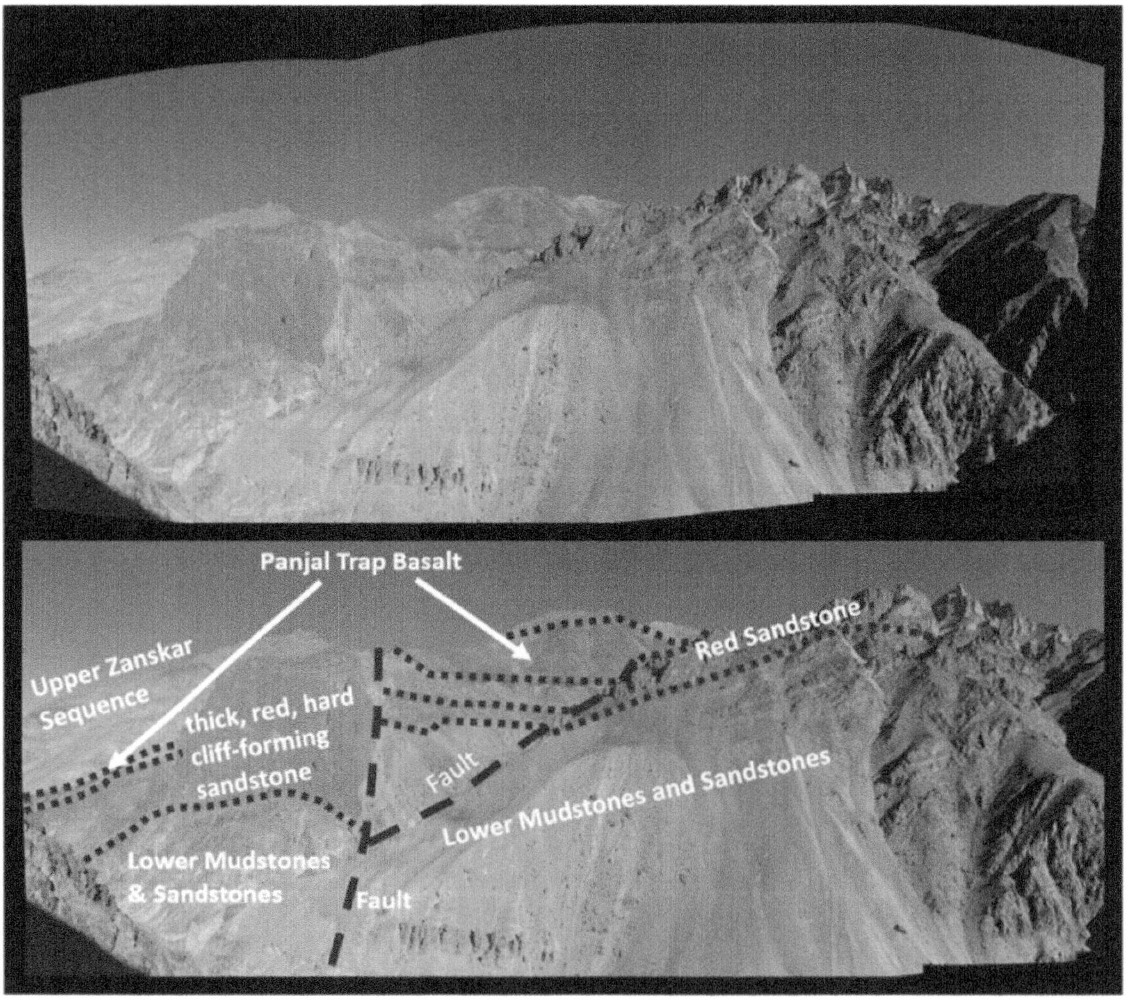

FIGURE 4.32 Photo-mosaic of outcrops in the Phuktal Valley. *Source*: Author's materials. Note: A lower geological interpretation of the photo-mosaic accompanies the upper panorama.

FIGURE 4.33 Chevron Folded Lower Zanskar Siltstones. Phe, Zanskar. *Source*: Author's Materials.

often used in road and highway surface markings) is common within the Lower Zanskar Sequence. Medium-thick-bedded siltstones, sandstones, and mudstones are folded within a chevron-folded *synform*. The fold is bisected by a *fold axial plane* which strikes along a compass bearing of 116⁰ (east-southeast) and dips at an angle of 44⁰ to the north-northeast. The *strike direction* of a plane is the line of zero dip. The maximum *dip direction* of a plane is at 90⁰ to the strike. Strike and dip directions are the most common structural measurements in field geology: they provide a three-dimensional geographical context for dipping planes such as bedding, schistosity, or foliation. Folds have two fold *limbs* on either side of the *fold axial plane*. Fold limb 1 dips to the north at a gentle angle of around 25⁰, whilst fold limb 2 dips steeply to the north at an angle of 70–80⁰. Folds provide a three-dimensional data frame for the paleo-stress fields that produced the fold structures. To produce the fold in Figure 2.39, with its geographical orientation, maximum compression must have been perpendicular to the strike of the axial plane: i.e., NNE-WSW. Maximum extension is parallel to the axial plane strike (ESE-WSW), and the intermediate stress is parallel to the axial planar dip. Note how the bedding thicknesses in fold limb 1 are significantly greater than bedding thicknesses in fold limb 2: limb thickening and thinning are produced within the shear stresses induced at the time of fold formation (see Marshak, 2019, chapter 11, for more details re folding).

Figure 4.34 is from the Phuktal area, Zanskar, and presents a classic example of an *unconformity*. The geological concept of an 'unconformity' was first proposed by a Scottish doctor called James Hutton, who saw similar structures in southeast Scotland during the 18th century. He rightly interpreted an 'unconformity' as marking a significant time gap between upper and lower rock units, on either side of the plane of unconformity. He went on to speculate that geology suggested 'no vestige of a beginning' in terms of time

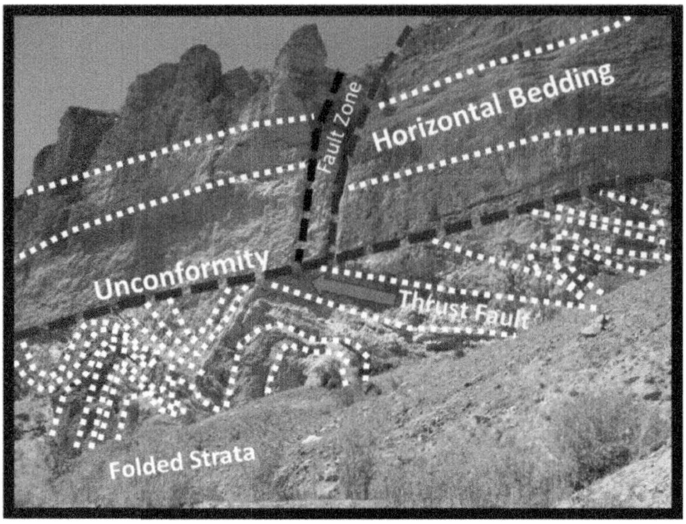

FIGURE 4.34 Upper horizontally bedded limestones and sandstones unconformably overlie folded older strata. *Source*: Author's materials.

(Hutton, 1788). In other words, the then-limited human concepts of Earth history timeframes were unrealistic. Earth requires millions and billions, not thousands, of years to account for the evidence present in the rocks. Figure 4.34 is an image of two parts: a lower older unit that has been folded and thrust-faulted through tectonics, and subsequently uplifted to the Earth's surface, where it was eroded. The upper unit represents a younger stage of sedimentary deposition from rivers that flowed through a renewed, uplifted, and eroded landscape. How long this unconformity represents remains subject to research, but probably requires millions of years between the tectonic forces that produced the folding in the lower unit (probably during the Himalayan Orogeny), and sedimentary processes that belong to a much more recent geological period. Folding is predominantly of a chevron style, although it becomes more complex through shearing and thrust faulting. The upper unit comprises weakly cemented horizontally or gently dipping conglomerates and sandstones. A fault zone has affected the upper but not the lower geology and is part of a younger period of tectonics; the Himalayas are an ever-dynamic geological environment.

Figure 4.35 presents a photograph on the right-hand side and a geological interpretation on the left-hand side of a complexly deformed rock sequence. The Himalayas, like other parts of the Earth's crust that have been twisted, torn, pushed, pulled, and squeezed by enormous tectonic forces, are a world-class laboratory to study the effects of crustal deformation. The uninterpreted image displays a wide range of structures and *structural disconformities* (planes that separate areas of crust which have structural integrity within themselves but are difficult to match with their neighbouring parts of geology). Disconformities in the case of Figure 4.35, shown in thick black continuous lines, are brittle faults that have moved crustal blocks around and caused intense shear adjacent to the faults. Lighter- and darker-coloured strata aid interpretation, as the colours represent different geological units. There is no obvious immediate structural continuity between the lighter and darker coloured geological units. A fault, probably a thrust fault, has emplaced the lighter coloured rocks on top of the darker coloured rocks. The separating fault between the lighter and darker coloured rocks is the primary structure; other faults are rooted in the primary faults but transmit stress, shear, and movement in different directions. The upper part of the fault architecture resembles the cross-section of a tree or flower: geologists call this shape a *flower structure*. The whole outcrop in Figure 4.35 is broken up by faults into discrete *fault blocks,* bounded by faults on all sides. Individual fault blocks comprise rocks which are folded and sheared. Chevron folds are common: chevron folds that point upwards are termed *antiforms,* and those that point downwards are termed *synforms*. There are many other fold shapes present in this outcrop. Some are highly contorted and 'broken', with

FIGURE 4.35 Complexly deformed Lower Zanskar Sequence sandstones and siltstones, Karsha Valley, Zanskar. *Source*: Author's materials. Note: The left-hand side diagram shows the orientation of bedding (white checked lines) and faults (red continuous lines).

some parts of folds detached from others: this results from high-intensity deformation within this piece of crust, creating chaotic fold shapes and architectures. The upper left-hand side of the photograph displays an *overturned fold*, in this case an overturned, upturned, *antiform*. Overturned folds have axial planes that deviate significantly from the vertical and include one limb that has turned 'upside down' from its original disposition. They are overturned through directional stress, and these can be used, along with other indicators, to model the palaeo-stress /shear direction at the time of fold and thrust formation. In this case, the shear stress is from the left to the right, as thrust movements also indicate: local deformation here involved shear stress and thrust faulting to the southwest (which is a common thrust direction in this area). Figure 4.35 structures were produced by SW-directed tectonics within the upper crust. As folding progressed, it could not accommodate the enormous compressive stress alone. Folds broke along '*detachment*' structures, or thrust faults, that detached one part of the crust from another, and moved it towards the southwest. Whilst much of the deformation was within the brittle state, local heating created by kinetic frictional forces of fault movements produced localised ductile deformation, allowing shapes to change and alter significantly.

LOWER ZANSKAR SEQUENCE LANDSCAPES

There are some strong topographic-forming units within the Lower Zanskar Sequence, particularly the hard red sandstones and quartzites. Much of the Lower Zanskar Sequence comprises weaker units with an abundance of mudrocks, siltstones, shales, and slates. These lithologies tend to form 'softer' topographies.

Figure 4.36 is a good example of typical topography within this geology. Mountains within this panorama are more rounded, with gentler slopes. Locally, there are examples of escarpments and dip and scarp slopes, reflecting geological structures. Several valleys have been cut into the ridge, which are fan-shaped in plan-view: narrow upper parts to valleys that pan out in a triangular form as they approach the plains. Continuous ridges characterise this topography, as opposed to ridges broken up by numerous individual high pyramidal peaks within the Higher Himalaya Crystallines landscapes.

Figure 4.37 illustrates the higher part of a ridge within the Tsarap Valley. Roads are cut into the ridges on both sides of the gorge-like valley. The shape of this ridge retains the rounder character of Figure 4.36, but also demonstrates how different parts of the Lower Zanskar Sequence stratigraphy create a range of micro topographies. Lower parts of the ridge are dominated by dark mudrocks with interbedded thin sandstones and limestones. These rocks produce extensive dark coloured screes that dominate much of the lower slopes. The absence of large cliff faces is characteristic of this lithology. Upper slopes provide quite a contrast to lower slopes. Thick red quartz-sandstones form large sheer cliffs, with strong scarp and dip slopes. These cliffs and crags can be followed for long distances along strike. Red quartz-sandstones produce sharp skylines, with deep clefts cut into the cliff faces. Scree-rich slopes contain light coloured stones and boulders, which contrast strongly with the darker scree emanating from mudstones. The regional dip of the rocks is apparent from this figure (dipping towards the left-hand side of the figure), which in turn shapes the form of the ridge. Bands of alternately coloured dark red and light coloured strata can be seen to lower in altitude from the right-hand side of the figure to the left-hand side. This indicates the rocks dipping to the left at 20° to 30° (towards the north). Roads themselves produce their own human-made landscape. They are an engineering feat,

FIGURE 4.36 Photomosaic of the southern ridge from Karsha with the Padum Plain in the foreground. *Source*: Author's materials. Note: This ridge is typical of the Lower Zanskar Sequence Mountain landscapes.

FIGURE 4.37 Mountains adjacent to the Tsarap Valley, close to Phuktal, Zanskar. *Source*: Author's photographic mosaic. Note: The difference in mountain shape between the lower, darker, softer slopes, composed of mudstones and siltstones, and the upper-left prominent red quartz sandstone unit which has produced harsher, more jagged cliffs and escarpment features.

resulting in armies of people alongside mechanical diggers and dynamite, blasting steep cuttings into the mountains. Roads in mountains such as the Zanskar range require regular maintenance. Landslips are a common phenomenon.

Figure 4.38 presents a range of topographies within the Lower Zanskar Sequence rocks. The River Tsarap forms the lowermost part of the local landscape. The Tsarap takes advantage of NNE-trending faults and geological structures, allowing the river to find its optimal route. As all rivers do, the Tsarap twists and bends through meanders that cut into rocks and sediments: Figure 4.38 presents the curve of the Tsarap River as it meanders around thick Recent alluvial deposits. The steep nature of the cliffs in the foreground testifies to the rapid downcutting power of the Tsarap River. A 'V' shaped valley occupies the upper centre of Figure 4.38. 'V' shaped valleys are typical of river erosion and rapid river downcutting. The geology dips steeply from right to left (to the north), forming strong dip slopes. A range of flat terraces in the landscape suggests that the downcutting of the river was halted for significant time periods: the rise of the mountains is not a continuous process. Tectonics will elevate mountains rapidly under certain conditions and halt elevation at other times.

Figure 4.39 beautifully illustrates the trap-like topography of this part of the Lower Zanskar Sequence stratigraphy, dominated by thick red quartz sandstones. The latter part of the day catches the red colour of the rocks, which dip downstream, to the north. Elevations of the lower contact of red sandstones decrease downstream. The turquoise blue colour of the river contrasts with the ochre, red, and brown colours of the rocks. The steep 'V' shaped river gorge evidences mountain uplift and river rejuvenation. The distant Upper Zanskar Sequence and Panjal Traps rocks are the subject of the next section of this chapter.

FIGURE 4.38 The River Tsarap meanders around a high ridge created from previous alluvial deposits, testifying to rapid river downcutting in response to tectonic uplift. *Source*: Author's photograph. Note: The side valley located within the upper backdrop is perched high above the Tsarap gorge, evidencing previous episodes of rapid downcutting. Phuktal. Zanskar.

FIGURE 4.39 Tsarap River Gorge, Phuktal, Zanskar. Note: Late-day sun catches the vivid colour of the red-quartz-sandstone, dipping downstream, towards the north, forming steep, craggy, valley sides. The deep river gorge occupied by the turquoise blue Tsarap River evidences rapid downcutting.

LITHOLOGIES, LANDSCAPES, AND FIELD GEOLOGY OF THE PERMIAN PANJAL TRAP BASALTS

Figures 4.40 and 4.41 present clear examples of the wonder of geological mapping. Geological maps, such as Figure 4.1, present geological data in a spatial form. Geological maps illustrate relationships between the geographical distribution of geological units (called Formations, Groups, Members, Sequences and so forth) and topography. Geological maps also convey the relationship between geological units: e.g., conformable stratigraphic contacts (that is, one unit continuously following from a younger underlying rock formation without a discontinuity) or faulted/unconformable contacts that indicate gaps in geological continuity. Geological mapping, as a process, involves geologists walking the ground, recording a wide range of geological and spatial information, examining remotely sensed images (satellite imagery and similar) and collating/synthesising all appropriate data to produce a geological map. Maps are interpretations and models, and change with advancing databases. Marker horizons such as the Panjal Trap Basalts are prominent, distinguishable, identifiable, and contiguous across landscapes: as such, they are a gift to geological mappers. Figures 4.40 and 4.41 illustrate how the Panjal Trap 'marker unit' tracks across landscapes over many kilometres, and helps geologists unravel the geological jigsaw puzzle. Figure 4.40 (upper photograph) tracks Panjal Trap Basalts from mountains above Tungri, through Karsha, to the northern Zanskar Valley, north of Stondey. The lower photograph in Figure 4.40 focuses on the Zanskar Valley area, north of Stondey. The marker horizon divides the stratigraphy into the Lower and Upper Zanskar Sequences. Figure 4.41 is from Stondey itself, where the contact between the Panjal Trap Basalts and the Upper Zanskar Sequence is beautifully exposed. Landscape analysis is a powerful way of examining geological structure and stratigraphy, helping to build geological maps and models. The very first geological map

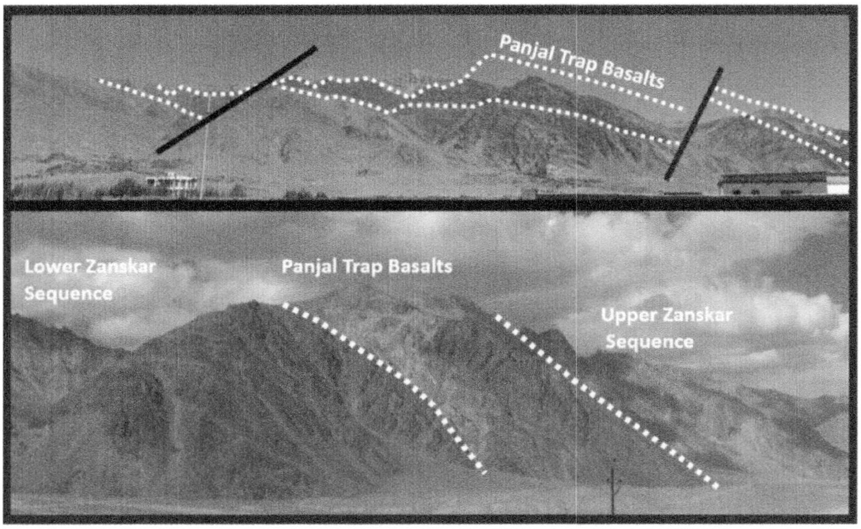

FIGURE 4.40 Permian Panjal Trap Basalts tracking across the Zanskar landscapes. *Source*: Author's materials. Note: Panjal Basalts form a stand-out prominent geological marker, separating Upper and Lower Zanskar Sequence rocks. The upper photograph indicates how the unit can be mapped for many kilometres across Zanskar mountains. The lower photograph presents Panjal Trap Basalts as sandwiched between Lower and Upper Zanskar Sequence stratigraphies.

FIGURE 4.41 Stondey Region, Zanskar. *Source*: Author's materials. Note: Example of how geologists use landscape morphology and geology together to map contacts between geological units and across landscapes. The left-hand side presents a strong escarpment formed by the Panjal Trap Basalts and the upper surface dipping across the valley. The right-hand side photograph indicates how the contact between the Panjal Basalts and Upper Zanskar Sequence rocks is faulted.

in a modern context was published in 1815 by an English canal engineer named William Smith (Smith, 1815). Through his many years overseeing the building of canals throughout Britain, Smith saw patterns in strata, fossils, and geology. This inspired him to record his observations in the form of a map. Smith's original map can be viewed at the Geological Society of London, Piccadilly, London, and viewed online. The map is remarkable in its overall accuracy: to this day, the map stands the test of time, although many modifications have resulted from new data acquisitions and interpretations. Simon Winchester authored a book entitled *The Map That Changed the World*, which describes in detail the life of William Smith and the impact of the map on the world (Winchester, 2001). Geological maps have indeed changed the world from philosophical perspectives towards space and time, to practical applications of geological knowledge. Maps are practical documents inspiring investigations for geo-materials such as metals, industrial minerals, construction minerals, aggregates, energy-minerals, and water resources, lending assistance to land-planning decisions, providing a base for hazard analysis and disaster risk reduction studies, and being invaluable for numerous types of engineering and infrastructure activities. They also form a serious basis for academic and scientific study.

Figure 4.42 presents aspects of the Panjal Trap basalts at the rock face in Stondey. Pecked white lines aid observer analysis of Figure 4.42. Basalts are by far the most common volcanic rock on Earth and represent the melts of Earth's mantle (Marshak, 2019, chapters 6 and 9). Basalts are erupted at temperatures above 900-1,100°C and can flow at lower or higher levels of viscosity depending on their eruptive temperature, chemical composition, and the presence of fluids within the lava. If they are erupted at very high temperatures (in the geological early history, some lavas were erupted at c. 2,000°C), their viscosity is extremely low. As they cool, or if erupted at lower temperatures, they flow more sluggishly. Figure 4.42 exhibits the presence of flow folds with a 'ropey-like texture'. The term for this texture is *pahoehoe,* which is a Hawaiian word for 'rope-like' and forms as low-viscosity 'runny' lava cools, with rope-like bands forming as the surface of the lava cools: the nature and orientation of the pahoehoe structures are controlled by the local flow behaviour of the lava.

Figure 4.43 presents vesicular (left-hand side) and porphyritic (right-hand side) volcanic rock textures. *Vesicles* are holes in the rock. They are formed through gas escaping from solidifying lava and can produce an aerated aero-bubble-chocolate-like texture. A *porphyritic texture* refers to the presence of larger crystals set within a finer-grained groundmass (background majority of the rock). The larger crystals are termed *phenocrysts* and record the previous growth of larger crystals taking time to grow within subterranean magma chambers. In this case, the white square crystals are feldspar and the dark crystals (square to rectangular) were originally pyroxene (altered through later metamorphism). The groundmass is fine-grained as a result of rapid cooling when erupted.

Figures 4.44 and 4.45 record another story for the Panjal Traps. The primary (original) character of basalts is dark grey in colour, with fresh igneous minerals and igneous textures. In Zanskar, the Panjal Trap basalts have been altered through metamorphism. They may have experienced more than one episode of metamorphism, but the Himalayan Orogeny is the most likely cause of metamorphism within the Panjal Trap basalts. Figure 4.44 depicts a range in colours of the basalt within one outcrop: from dark grey to

FIGURE 4.42 Flow folding in the Panjal Trap Basalts, Stondey, Zanskar. *Source*: Author's materials. Note: Pecked lines outline flow folds within a low-viscosity basalt lava.

FIGURE 4.43 Vesicular (left) and porphyritic basalt textures (right), Stondey, Zanskar. *Source*: Author's photographs. Note: Vesicles form from the degassing of magmas/lavas. A porphyritic texture is one with larger crystals set in a fine-grained groundmass. Pencil in the left hand side image is 15cm long. 4cm of the pencil is exposed within the right hand image.

FIGURE 4.44 Greenschist metamorphism of Panjal Trap Basalts. *Source*: Author's photograph. Note: The strong green colour of the metamorphic mineral epidote, amidst less-altered grey-black basalt. Stondey, Zanskar.

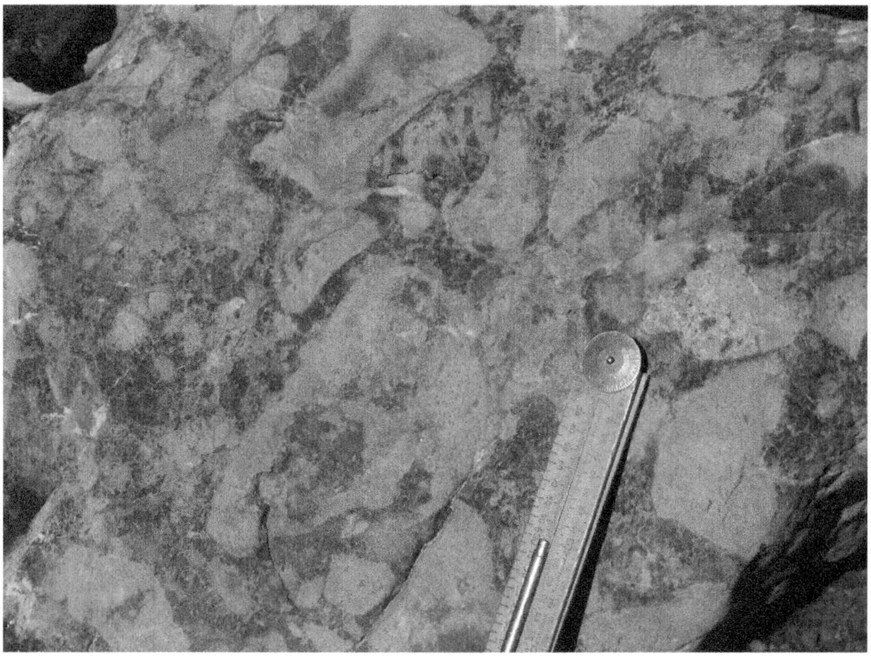

FIGURE 4.45 Brecciated basalt, altered to epidote. Karsha, Zanskar. *Source*: Author's photograph.

red and green. The dark grey colour reflects the less altered nature of the basalts. Other colours indicate metamorphic processes altering the primary nature of the rock. In this case, the basalt has been metamorphosed to greenschist grade, with prominent pistachio-green epidote present. Metamorphism is often accompanied and facilitated by high volumes of hydrothermal fluids, which are transmitted through the rock mass at elevated temperatures, greatly assisting mineral and chemical reactions to occur. The basalt presented in figures 4.44 and 4.45 was subject to significant chemical and mineralogical change. Original pyroxene-feldspar dominated mineralogy changed to epidote and chlorite, which are visible as green alteration zones within Figure 4.44. Careful observation of Figure 4.44 reveals the diffuse nature of the alteration process, leading to variable degrees of change within the rock-mass. Figure 4.45 depicts a *breccia*: a rock composed of angular-sub-rounded large particles (clasts) and smaller clasts set within a finer grained matrix (silt-grade). Large clasts have been transformed to the green metamorphic mineral epidote, whilst the finer-grained matrix remains dark grey and is less obviously altered. *Conglomerates and breccias* (conglomerates with rounded large clasts) are poorly sorted rocks with a wide range of *clast* (rock fragment) sizes: in this case, larger clasts extend to up to 15 cm, situated within a matrix of mm-sub-mm clasts. Volcanism can produce many different types of rock. Quieter *effusive volcanism* produces lavas. More violent gas-charged eruptions produce *pyroclastic rocks* (volcanic rocks composed of solid-clast particles as opposed to lava-liquid originated crystalline rocks). Each clast is a volcanic crystalline rock which solidifies prior to eruption. Pyroclastic rocks form from airborne eruptions, and eruptive products which flow along the ground and are composed of solid particles and gases (not liquids): the latter are termed *pyroclastic flows*. Post eruption, the hot clasts weld together to form pyroclastic rocks (Marshak, 2019, chapter 9).

To understand the context of the Panjal Trap basalts, it is helpful to understand a little about how the basalts were formed and their significance. Panjal Trap basalts have been dated by radiometric dating methods, yielding an age of 288 Ma, which is Lower Permian. At this time, India was part of the supercontinent Gondwanaland within the southern latitudes of Earth, connected to Africa, Australia, and Antarctica. *Extensional tectonics* (a style of tectonics that forms rift valleys and can split continents apart) and mantle diapiric dynamics in the form of balloon/mushroom-shaped high-temperature plumes (plumes are anomalously hot volumes of mantle which can rise through the Earth's mantle due to their lower density, termed *diapirism*), split Gondwanaland, detaching some fragments from northern India and surrounding parts of Gondwanaland. A new ocean called Tethys formed at this time. This continental break-up involved the rapid eruption of large-volume basalts, some of which are preserved in India and Pakistan. In Kashmir, these plume-related volcanic rocks reach up to 3 km in thickness. Although basalts dominate the sequence, other volcanic rocks, such as andesite and dacite, occur. In Zanskar, the basalts are mostly tens of metres to c. 100 m thick. For further details, see Bhat and Zainuddin (1978), Chauvet et al. (2008), and Shellnutt et al. (2011, 2023).

LITHOLOGIES, LANDSCAPES, AND FIELD GEOLOGY OF THE TRIASSIC-EOCENE UPPER ZANSKAR SEQUENCE

Figure 4.46 presents a typical landscape comprised of Upper Zanskar Sequence limestones and calcareous sandstones/siltstones. The Lingshed mountains expose folded thinly to moderately-thickly bedded and folded blue-grey, reddish brown, yellow, and buff-white coloured strata. Boundaries between individual units can be followed across the landscape.

The Upper Zanskar Sequence is Triassic-Eocene in age. The bulk of the sequence is Mesozoic (Triassic, Jurassic, and Cretaceous, aged between 251 and 66 Ma, or almost 200 million years of time) and all sequences presented in this book are Mesozoic in age. There is a marked change in character of the sedimentary rocks between the Lower and Upper Zanskar Sequences. The Upper Zanskar Sequence is characterised by lighter-coloured, thin to moderately and thickly bedded calcareous siltstones, sandstones, and limestones with non-calcareous mudstones and siltstones. As noted previously, this book does not focus on stratigraphic details of the Zanskar Sequence, except for differentiating it into Lower and Upper units. The Upper Zanskar Sequence is divided into at least 12 stratigraphic units, including the Kuling Formation, the Quartzite Series, the Kioto Limestone Group, the Giumal Sandstone, and the Kangi La and Chulung La Formations. Lithologically, the Upper Zanskar Sequence comprises many types of limestones, shales, mudstones, and siltstones, together with a variety of sandstones. Sedimentary facies interpretations of much of the Lower and Middle Mesozoic Zanskar Sequence revolve around stable shelf/shallow marine settings. As the Indian Plate approached Eurasia, the stable shelf environment evolved into deeper marine basin environments, alongside fluvio-deltaic and flysch settings. For further reading, see references already given in the Lower Zanskar Sequence section, together with Fuchs, 1987, Baud et al., 1984, and Garzanti, 2024.

FIGURE 4.46 Photo-mosaic panorama of carbonate-rich Upper Zanskar Sequence bedded limestones and sandstones. Lingshed, Zanskar. *Source*: Author's photomosaic.

Figure 4.47 presents three lithological units within the Upper Zanskar Sequence. Rocks dip to the right-hand side of the photograph and strike along the ridge for many kilometres. Contacts between the three units are visible and can be tracked across the landscape. The blue-grey colour of the lowermost unit is typical of limestone lithologies, the central darker unit contains a higher proportion of mudstone and siltstone, whilst the uppermost yellow-brown-buff unit is comprised of more thinly bedded calcareous siltstones and sandstones. Brightly coloured scree drapes over the lower, darker coloured strata, washed down by mountain streams in times of rainfall. Figure 4.48 illustrates a typical outcrop-scale example of well-bedded Upper Zanskar Sequence limestones. Bedding thicknesses here mostly vary between 30 cm and 2 m. Dissolution structures, typical of limestones, are particularly evident within the central beds of Figure 4.48.

Bedding represents a time-gap in sedimentary deposition and is structurally highly valuable to geologists, as it provides a key tool to understand structure. Bedding is a primary horizontal structure. Tectonics and deformation change the orientation of bedding, causing beds to dip (tilt) relative to the horizontal. Dip measurements are basic quantitative data that field geologists use to understand the structure of an area. Dips can vary from low angles (c. 10° to the horizontal) to moderate (c. 30–50°) to vertical attitudes (e.g. Figure 4.48). They can also be overturned and become 'upside down' relative to their original, primary orientation. Figure 4.49 presents vertical to steeply dipping beds. Figure 4.49 (left side) illustrates thin- to moderately-thickly bedded limestones and sandstones, whilst Figure 4.49 (right side) presents very thickly bedded limestones adjacent to thin-moderately thickly bedded sandstones and limestones. Very-thickly

FIGURE 4.47 Bedded strata of the Upper Zanskar Sequence extend across kilometres of terrain. *Source*: Author's photograph. Note: Three units are clearly visible: lowermost blue-grey limestone, a central darker shale and limestone, and upper yellow-brown calcareous sandstones and siltstones. Zanskar Valley.

FIGURE 4.48 Thin-moderately, vertically, bedded limestones and calcareous sandstones. Rinam, Zanskar. *Source*: Author's photograph. A c. 10m long x 5m high outcrop is presented in Figure 4.48.

FIGURE 4.49 Thick-bedded limestones (right side) and more thinly-bedded calcareous sandstones (left side). Rinam. Zanskar. *Source*: Author's photographs.

bedded limestones are produced through protracted periods of continuous sedimentation, whilst thinly bedded sandstones and limestones indicate more rapid variations and changes in sedimentation. Much of the Mesozoic Upper Zanskar Sequence was formed in relatively shallow marine conditions. Limestones form from chemical precipitation of calcium carbonate, most effectively through marine organic processes. Thick limestones mainly form within warm to tropical shallow marine and relatively quiet seas with high organic productivity. Molluscs such as bivalves, gastropods, and brachiopods thrive in these sediment-free, warm, shallow marine environments, growing abundant hard-calcareous shells (Marshak, 2019, Chapter 7). An excess of clastic sediment within shallow marine environments, in the form of sand and silt, can choke marine organisms. Rock units presented in Figure 4.49 indicate a variation in palaeo-environments from non-clastic (sedimentary rocks that form mainly through chemical precipitation), sediment-free carbonate reef conditions, to one of shallow to moderately deep marine conditions, experiencing regular sediment-charged currents, depositing bedded sandstones and siltstones.

Figure 4.50 is a close-up of c. 1 m-thick beds of homogeneous blue limestone with thin interbeds of brown mud/siltstone. This limestone outcrop has a crystalline texture, varying between coarse and fine grained: geologists term these limestone textures *sparite* and *micrite* (coarse- and fine-grained crystalline limestone).

Figure 4.51 illustrates the incorporation of mud flakes within limestones. Mud flakes here are up to 12 cm × 1–2 cm in length/breadth and are aligned parallel to the bedding. Mud flakes indicate the co-existence of mud clasts (particles) within the predominant carbonate environment. Mud particles gradually filled hollows and interstices within the carbonate reef; there were times when mud dominated and covered the reef. Mud is an indicator of quiet water conditions.

Figure 4.52 is a close-up of more thinly bedded calcareous siltstones and fine-grained sandstones. Colours of bedded units vary from pale grey to blue grey, and vary in thickness from a few millimetres to 10 cm. Beds appear to be disrupted, with the pale grey finer-grained units displaying *'bioturbated' textures*. *Bioturbation* is a term that defines the physical movement and mixing of soil/sediment through the activity of organisms, such as burrowing worms. In this case, bioturbation occurred on a shallow marine seabed, with a substrate formed from calcareous muds and silts. Within the seabed, burrowing and boring organisms ingest, digest, and excrete substrate materials and intermix these materials with silts and sands. At the rock face, beds can be seen to wedge out laterally or become incorporated by other units. Larger clasts (particles) are visible that are several millimetres in diameter and are gravel clasts and mud flakes (see Marshak, 2019, chapter 7, for more details).

Lower Zanskar limestones and calcareous sandstones were subject to deformation, heating, dehydration, and low-grade metamorphism (as will be discussed in detail below). Figure 4.53 illustrates a dense vein network comprising calcite and silica veins, which were formed through the activity of hydrothermal

FIGURE 4.50 Metre-thick beds of blue crystalline limestone, with thin interbeds of brown siltstone and mudstone. Chazar. Zanskar. *Source*: Author's photograph. The photograph includes a 50cm length of a 1m ruler.

FIGURE 4.51 Blue limestones with elongate flakes of mudstone (greenish lenticular structures within the body of the limestone). Chazar. Zanskar. *Source*: Author's photograph.

FIGURE 4.52 Centimetre-decimetre bedded bioturbated calcareous siltstones and sandstones. Rinam. Zanskar. *Source*: Author's photograph.

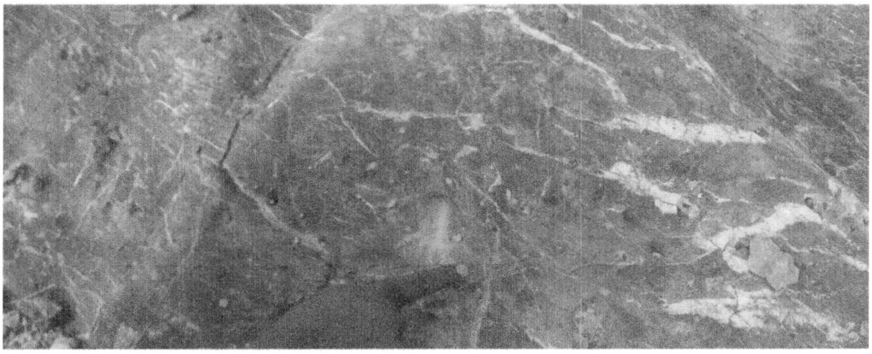

FIGURE 4.53 Veined limestone outcrop. Zangla. Zanskar. *Source*: Author's photograph.

solutions, with dissolved chemicals in solution. As local conditions changed, due to drops in pressure and temperature, hydrothermal solutions precipitated calcite-silica minerals within veins: veins filled cracks in the rock mass produced through deformation.

FOLDING AND FOLD-INFLUENCED LANDSCAPES WITHIN THE UPPER ZANSKAR SEQUENCE

Figure 4.54 presents a dramatic introduction to the complexity of folding within the Upper Zanskar Sequence. Folds presented in Figure 4.54 (within limestones and calcareous sandstones and siltstones) are *nappes*, defined as flat-lying folds, with horizontal axial planes. Nappes can transport material

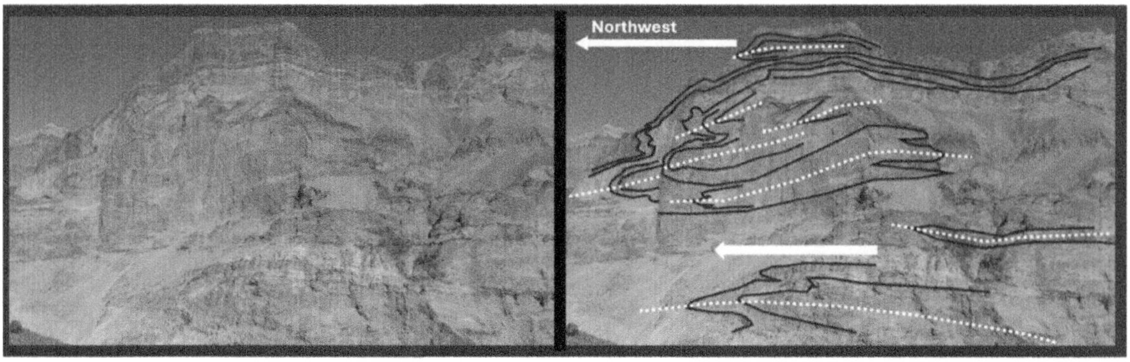

FIGURE 4.54 Mountain-scale flat-lying folds (nappes). *Source*: Author's materials. Note: Limestone and carbonate sandstones and siltstones have been extensively folded. Folds have horizontal axial planes. Compare the photograph on the left with the geologist's interpretation on the right. Sengge La. Zanskar.

sub-horizontally for considerable distances and thicken crust. Nappes and nappe processes are fundamental to mountain building, found in all compressive orogenic belts around the world. The right-hand side of Figure 4.54 gives a geologist's line interpretation of the folded strata. The left-hand side remains unmarked, allowing the reader to compare and contrast structures and interpretations. Folds close to the left-hand side of Figure 4.54. Fold limbs are flat lying and isoclinal (display parallel orientations around the centre of the fold). Black continuous lines follow individual strata, emphasising fold shapes. White pecked lines illustrate the track of fold axes: these are flat-lying, a characteristic of a nappe fold. At least four or five separate nappe structures are present within Figure 4.54, and indicate tectonic movement towards the left-hand side of the photograph (towards the northwest). Figure 4.54 exhibits a pile of nappes: nappes stacked one on top of another, clearly presenting crustal processes that thicken crust within orogenic zones.

The Upper Zanskar Sequence presents numerous examples of folding, tectonic movement, tectonic-vector-directions, and crustal thickening. Deformed bedded strata are particularly informative with respect to understanding and unravelling tectonic movement and crustal thickening processes. These structures influence the shape of the landscapes in Zanskar. Figure 4.54, for example, illustrates flat-topped ridges that reflect the flat-lying nature of nappes. The field area underlain by the Upper Zanskar Sequence is a showcase for the folding of bedded rocks at all scales: from microscopic to mountain scales. To understand the geometry of folds and how they are used by geologists to unravel the complexities of mountain building, deformation, and tectonics, it is instructive to briefly discuss the geometry and stress fields that produce folds.

Figure 4.55 presents the definitions of various parts of fold geometry. Folds have a similar shape to sine curves or water waves. Fold terminology is close to terms used for all types of waves. The wavelength, for example, is the distance between the same successive points of a wave or a fold. Folds that close upwards are termed *antiforms*. Folds that close downwards are termed *synforms*. Folds are bisected by an *axial plane*. Axial planes bisect folds in three dimensions.

Hinge zones are areas closest to the fold closure zones. The *fold axis* is the line of a synform or *antiform* crest. *Fold inflexion* points are halfway between antiformal and synformal hinge zones. *Limbs* form those parts of folds that join the hinge zones of antiforms and synforms. The angle between limbs is termed *interlimb angle*: folds are given different names depending on their interlimb angle (gentle between 180° and 120°, open between 120° and 70°, close between 70° and 30°, tight for <30°, and isoclinal for 0°). The amplitude of a fold refers to the half-distance between the synforms and antiforms' high/low points. Terms such as conical and cylindrical refer to the resemblance between fold geometries and those of either cones or cylinders, and replicate these types of geometries throughout their fold shapes. Some folds are non-cylindrical and non-conical, and have curved hinge lines and fold axial planes: many of the folds presented here are non-cylindrical. Folds form due to compression. In Figure 4.55, an example is given

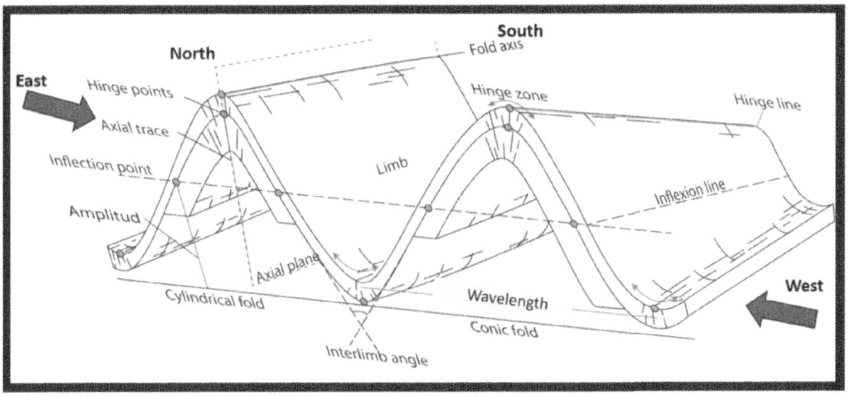

FIGURE 4.55 Fold geometries and definitions of fold components. *Source*: Full acknowledgement via Wikimedia to Alejandro GuevaroAlday (2019). https://commons.wikimedia.org/wiki/File:3D_Fold_diagram.jpg. Reference to Creative Commons licence CCA-SA 4.0 International. https://creativecommons.org/licenses/by-sa/4.0/deed.en. The original figure has been slightly modified by adding hypothetical geographical direction indicators (East and West) to allow a discussion to take place in the text.

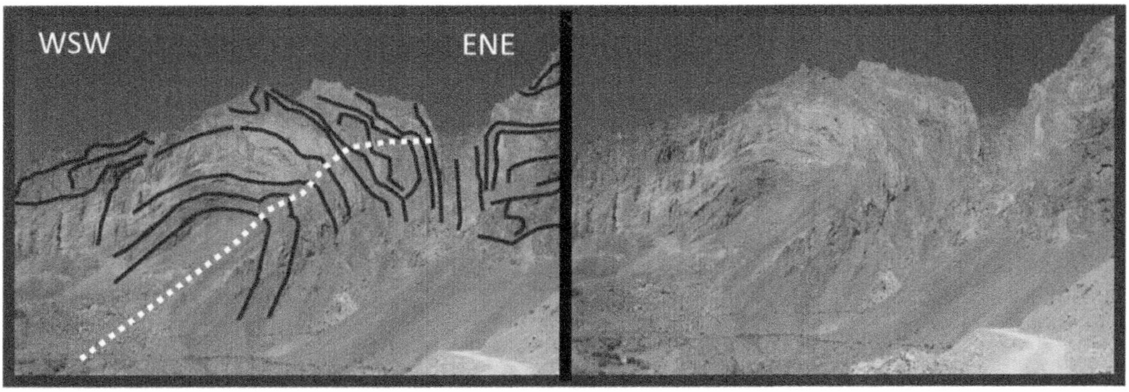

FIGURE 4.56 Mountain-scale folds, Lingshed, Zanskar. *Source*: Author's materials. Note: White checked lines delineate fold axes. Red dashed lines indicate faults. Black continuous lines follow bedding planes. The most prominent structure is an overturned antiform. Compare the geologist's interpretation on the left with the photograph on the right.

of folds which have axial planes that trend north-south and result from compression directed east to west. Mapping the geometry of folds provides data for the likely tectonics and compression-extension dynamics that existed within the mountain belts at the time of fold formation.

Figure 4.56 presents a remarkable outcrop of folded yellow, brown, and buff calcareous strata around Lingshed. Around 1,000 m of continuous vertical exposure presents opportunities for the study of large-scale folding. The main fold structure on the left-hand side of the photographs is an overturned antiformal fold, closing upwards. The left-hand side photograph provides a geological interpretation. Black continuous lines depict bedding orientations, the white pecked line depicts the surface trace of the fold axial plane, and red dashed lines indicate faults/fractures. West-southwest (WSW)-to east-southeast (ENE) compressive stress produced this fold. The axial plane dips to the WSW, indicating local tectonic movement towards the ENE. Mountain shapes are profoundly influenced by fold geometries. The curvilinear nature of the large fold is reflected in the curvilinear shape of the mountain crests. Gullies and valleys tend to form along steep fold limbs, and within fractures and faults, which in turn are produced through shearing of folds. Fold processes alone could not accommodate the amount of shortening local tectonic forces demanded. Fold structures fracture and fault when shortening exceeds the capacity of folding to accommodate the magnitude of crustal compression alone. Shearing of folds breaks up the continuity of fold architecture as can be seen in Figure 4.56: the antiform on the left-hand side would have produced a synformal fold that closed downwards on the right-hand side of the photograph. This structure is absent, suggesting that shearing and faulting have broken and detached the original structural integrity.

Figure 4.57 presents a whole valley and mountain scene with a vertical distance of over 1,500 m between the valley floor and the highest mountains. All mountains present in Figure 4.57 are fold mountains: mountains intimately influenced in their genesis and subsequent landscape formation by fold structures. The inset within Figure 4.57 provides a geological interpretation of the highest ridge, oriented from south-southwest (SSW) to north-northeast (NNE). The ridge face presents a series of northeast (NE) facing cliffs and steep slopes. These different views of the fold structures enable a detailed analysis of the three-dimensional nature of fold structures to be undertaken. Black lines indicate the orientation of key strata and their bedding structures. A range of fold shapes is exposed. The most common fold shapes are zig-zag chevron style folds with pointed hinge zones. Antiforms (upward closing) and synforms (downward closing) chevron folds are visible. Folding within the main body of the ridge is more curvilinear in shape, with most of the length of the ridge formed by a long upward closing curvilinear antiformal fold structure. Pecked white lines indicate the surface position of fold axial planes. These run along the NNE-SSW trending ridge and are exposed 'edge-on' along the NE facing cliffs and steep ridge faces at the right-hand side of the main ridge in the photograph. The larger fold structure 'plunges' towards the SSW: the surface hinge line of the crest of the antiform dips/plunges towards the SSW. The larger fold structure was produced by local WNW-ESE compression. There are numerous smaller-scale fold structures present alongside the larger-scale fold structure.

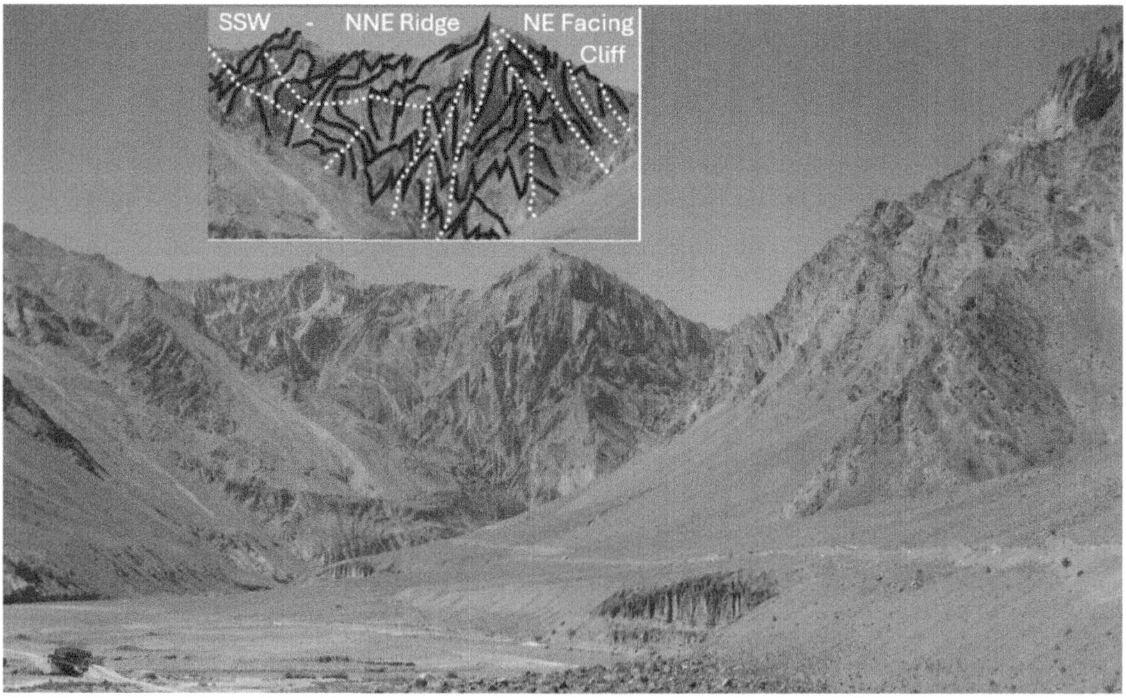

FIGURE 4.57 View from the road NW of Zangla, illustrating mountain-scale folds which shape the mountains themselves. *Source*: Author's materials. Note: Inset displays bedding orientations (black lines) and fold axial planes (white checked lines).

Folds occur at a wide range of scales, from microscopic to mountain scale. Geologists classify the scale of fold structures into major and minor structures in the field. Often, the larger-scale structures are unexposed: in Ladakh, the desert mountain environment allows for whole-mountain outcrops to be visible. Most climates do not produce such a wealth of field outcrop. Rocks are mainly covered by soils, surficial deposits, and vegetation. Geologists piece a whole-regional geology together through discrete observations of individual outcrop locations that may be far apart. It is far more likely that minor fold structures will be observed, and these provide keys to understanding the larger scale structures. Figure 4.57 presents an ideal scenario where minor and major fold structures are visible together, and the relationships between the two scale structures can be analysed in depth.

Figure 4.58 illustrates well-exposed fold geometries north of Zangla. Figure 4.58 is a gorge outcrop with a vertical relief of c. 30 m from top to bottom of the photograph. These folds are classic chevron (zigzag) folds with pointed fold crest morphologies and wavelengths of c. 10–15 m. Folds have an interlimb angle of <30° classifying them as tight folds. Upturned antiforms and downturned synforms are clearly visible. These folds resulted from ENE-WSW tectonic compression. Shearing has compressed the fold geometries and disrupted folded integrity through faulting.

Figure 4.59 presents a clear example of a curvilinear antiform in three dimensions. Other folds will be present within the figure, with this particular fold being particularly standout. The geologist's interpretative inset in Figure 4.59 labels the various parts of the fold in terms of hinge line, axial plane, and limbs. This fold has an interlimb angle of 70º, classifying it as a close fold. The three-dimensional nature of folds manifests here as a lateral continuity to the fold structure within the mountain geography. The hinge zone /axial plane strike is orientated NW-SE. This fold was produced by NE-SW tectonic compression. The axial plane is overturned, dipping towards the northeast, suggesting that local crustal movement was towards the southwest.

Figure 4.60 is a photomosaic of a 3 km long ridge section between Stondey and Chazar, Zanskar. The figure presents complex fold geometries that constitute a whole ridge. 1,200 m of vertical relief is visible from the valley floor to the highest part of the ridge. Figure 4.60 displays the ridge itself in the upper

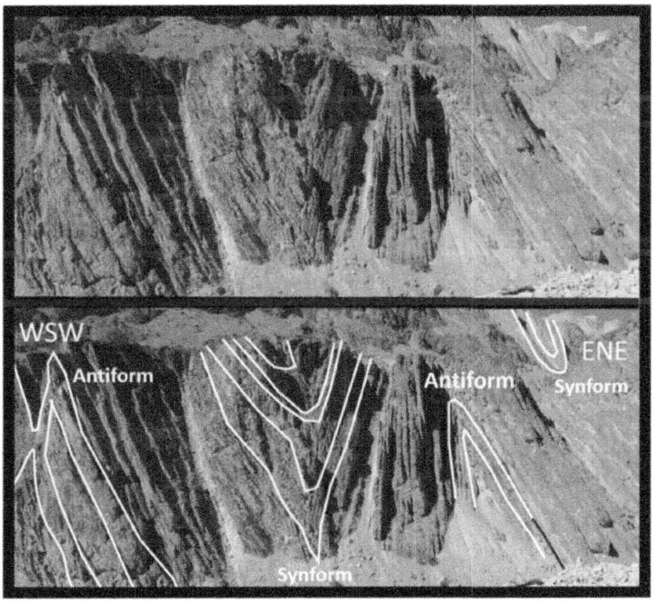

FIGURE 4.58 Zigzag (chevron) folds in the Zanskar gorge NW of Zanga. Zanskar. *Source*: Author's materials. Note: The lower figure presents a geologist's interpretation. White lines indicate bedding orientation.

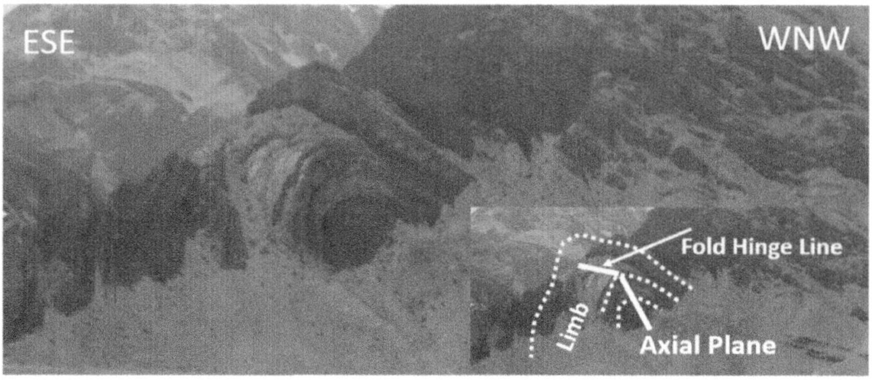

FIGURE 4.59 Well-exposed fold antiform in 3 dimensions, NW of Zangla, Zanskar. *Source*: Author's materials. Note: Inset provides a geologist's interpretation indicating bedding orientations and locations of the axial plane and fold hinge.

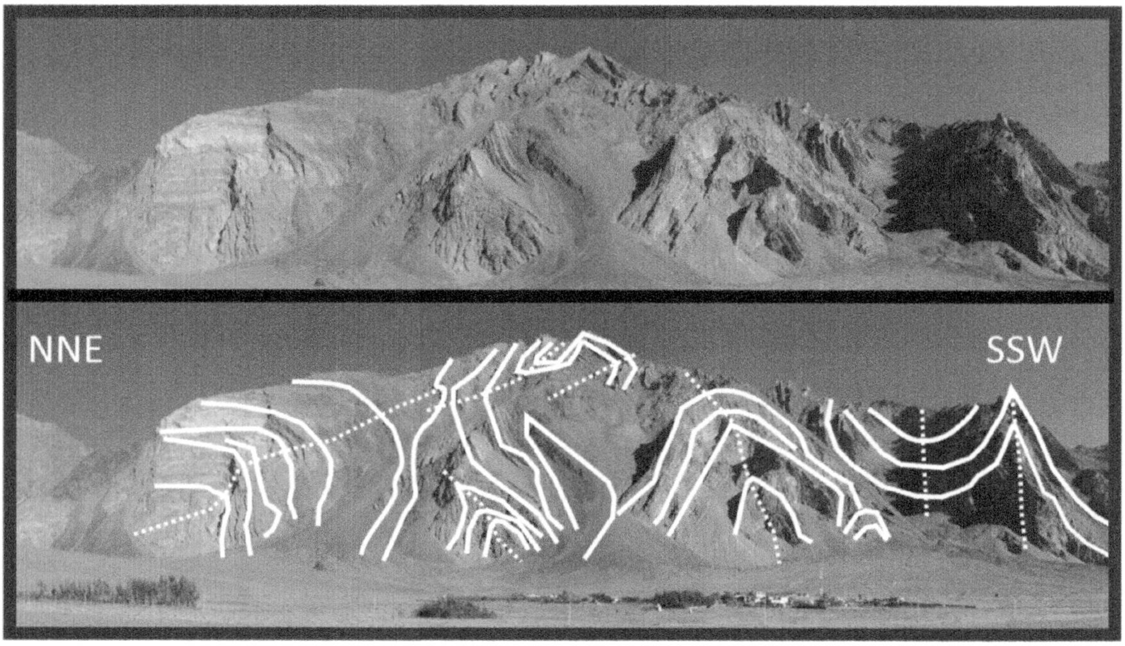

FIGURE 4.60 Fold structures shaping mountains between Stondey and Chazar, Zanskar. *Source*: Author's materials. Note: Solid white lines indicate bedding orientations. Checked white lines indicate fold axial planes. The red continuous line indicates the fault. A range of fold geometries is present, including overturned and chevron morphologies.

photograph. Geological interpretations are added in the lower photograph, including white continuous lines to indicate bedding orientations and a red line for a fault. The upper photograph clearly displays the shape of the ridge. The ridge crest has a central high point and lower flanks to the north-northeast (NNE) and south-southwest (SSW). The central peak has a triangular shape to it. The orientation of the NNE side of the ridge is strongly controlled by fold hinge lines and axial planes. The pointed nature of the SSW ridge end is shaped by the closure of a tight chevron-shaped antiform. Serrated skylines across the ridge are produced by vertical and steeply dipping strata. Two strong topographic features at the front of the ridge, facing the observer, are triangular-curvilinear shaped, dictated by the form of the folds that constitute these features. The three-dimensional shape of the ridge is a clear example of the close relationships between geology, geological structure, and topography.

Chevron and curvilinear folds are present in Figure 4.60. Three or four larger-scale folds are exposed. One fold dominates the ridge from the ridge peak to the NNE edge. This structure is a curvilinear fold with a fold hinge trace parallel or sub-parallel to the ridge crest orientation. The hemispherical form of the fold is expressed within the contours of the ridge crest zone. The fold axial plane dips towards NNE with a SSW-NNW strike: tectonic compression orientated ESE-WSW produced this fold and other folds of this orientation. The overturned nature of this fold suggested crustal movement occurred towards the WSW. This fold structure has an exposed length of c. 1,200 m.

Two folds with different orientations are present at the front of the ridge and towards the eastern side. These folds have axial planes oriented NW-SE, resulting from NE-SW tectonic compression. The scale of folds varies, with the largest folds having a fold height of c. 600 m and a wavelength of c. 1.5–2 km. Smaller scale folds have fold heights of c. 100–200 m and wavelengths of c. 3–500 m. Smaller-scale folds will exist but not be visible in the figure.

The fault (note the red line) on the SSW side of the ridge separates an antiform from a synform. These may have been originally joined, suggesting that the SSW side of the ridge has been pushed upwards towards the NNE. This situation is an example of a fold structure which could no longer accommodate shortening and 'broke'. The fold was detached and separated by a reverse fault that moved the fault block on the right-hand side (to the SSW) over the fault block on the NNE. Folding and faulting within compressive mountain belts are intimately related. Faults occur alongside folds within relatively cooler and more brittle environments.

Figure 4.61 presents such spectacular folds that they are even noted in Zanskar travel guides. The 'crazy' relationship between highly complex fold geometries and the shape of the ridges fascinates people with little or no geological knowledge: tourists and travellers enjoy visualising many different shapes in the mountain. Figure 4.61 presents a geological interpretation of the fold structures. Continuous black lines show the disposition of strata, whilst pecked white lines indicate the surface expression of fold axial planes. One thing is clear: the whole mountain is composed of folded strata and, as such, is a classic example of a 'fold mountain'. The vertical height between the valley floor and the highest mountains is around 1,400 m. Some 1.5–2 km horizontal distance is present in Figure 4.61. Perhaps the most eye-catching structures occur close to ground level. The ridge is separated by a deeply dissecting river valley (or tokpo in the local language). The left/SSW side of the lower ridge is dominated by a central antiform, which is approximately isoclinal (both limbs have the same angle to the hinge zone). Another antiform (with no intervening synform) is situated to the right of the large antiform. The axial trace and fold hinge line can be followed on the flat sand-covered ridge above these structures. An antiformal structure is present left of the central antiform: this could have been previously connected with the antiform, but is now detached, and thrust over the central antiform. The right-hand side of the ridge, to the NNE, east of the central river valley, is composed of a lower 'M' shaped fold, which is part of a larger overturned antiform occupying much of the NNE side of the ridge. The shape and orientation of this fold suggest local crustal movement towards the NNE. The remaining central and higher parts of the ridge comprise a series of folds of different shapes and sizes, with chevron folds being the predominant fold style.

Some folds present in Figure 4.61 are relatively simple, resembling a cylinder or part-cylinder (such as the central antiform on the SSW side of the river valley). Most other folds have complex geometries that change shape with increasing height within the fold structure. Folds can become *disharmonic* with the geometry not being 'in-harmony' with itself as fold geometries change over space. Axial planes can be curvilinear, indicating folds have become non-cylindrical.

FIGURE 4.61 Spectacular mountain-scale folds around Rinam, Zanskar. *Source*: Author's materials. Note: For geologists' interpretation: black lines indicate orientations of bedding, white checked lines indicate the orientation of fold axial planes. Some folds are cylindrical/conical and display simpler geometries. Others have curvilinear axial planes, indicating how the fold shape changes in space. Predominant NNE-SSW compression produced the folds, although a locally complex stress field reshaped and reorientated fold geometries.

Antiforms are adjacent to antiforms in parts of Figure 4.61, with no intervening synform, indicating that strain and stress have been accommodated not only by folding, but by shearing and faulting. Intervening synforms have been removed or 'faulted out' through the activity of faults and shears. Shearing within ductile environments breaks up fold structures. Different beds behave differently when folded: for example, a shale/muddy bed can be more easily thinned and stretched compared to a sandstone, which is more likely to break and fault. The variable behaviour of different materials can lead to the evolution of disharmonic, complex fold geometries.

This chapter ends looking southwards across the Indian Plate, having documented its constituent basement gneiss and cover sedimentary sequences. We now look northwards to the island arc terrane of Kohistan-Ladakh and the Karakoram Terrane.

REFERENCES

Baud, A., Gaetani, M., Garzanti, E., Fois, E., Nicora, A., & Tintori, A. (1984). Geological observations in southeastern Zanskar and adjacent Lahul area (northwestern Himalaya). *Eclogae Geologicae Helvetiae*, 77(1), 171–197. [Source 1.5.4]

Bhargava, O. N. (2015). Early Paleozoic paleogeography, basin configuration, paleoclimate and tectonics in the Indian Plate. *Journal of the Palaeontological Society of India*, 60(1), 71–86.

Bhat, M. I., & Zainuddin, S. M. (1978). Geochemistry of the Panjal Traps of Mount Kayol, Lidderwat, Pahalgam, Kashmir. *Journal of Geological Society of India*, 19(9), 403–410.

Butler, R. W. H. (2019). Tectonic evolution of the Himalayan syntaxes: The view from Nanga Parbat. In P. J. Treloar & M. P. Searle (Eds.), *Himalayan Tectonics: A Modern Synthesis*. Geological Society, London, Special Publications, 483, pp. 215–254. https://doi.org/10.1144/SP483.7.

Chauvet, F., Lapierre, H., Bosch, D., Guillot, S., Mascle, G., Vannay, J. C., ... & Keller, F. (2008). Geochemistry of the Panjal Traps basalts (NW Himalaya): Records of the Pangea Permian break-up. *Journal of Asian Earth Sciences*, 32(5–6), 332–349.

Fuchs, G. (1987). The geology of Southern Zanskar (Ladakh). *Jahrbuch der Geologischen Bundesanstalt*, 130(4), 465–491.

Gaetani, M., Casnedi, R., Fois, E., Garzanti, E., Jadoul, F., Nicora, A., & Tintori, A. (1986). Stratigraphy of the Tethys Himalaya in Zanskar, Ladakh. Initial report. *Rivista Italiana di Paleontologia e Stratigrafia*, 91(4), 443–478.

Gaetani, M., & Garzanti, E. (1991). Multicyclic progradational sequences in the Tethys Himalaya (Zanskar, India). *Geological Society of London, Special Publications*, 57, 171–187.

Garzanti, E. (2024). The stratigraphy of the Tethys Himalaya. *Italian Journal of Geosciences*, 143(3), 74–96.

Garzanti, E., Casnedi, R., & Jadoul, F. (1986a). Sedimentary history of a Tethyan continental margin (NW Hindukush–Zanskar Himalaya, India). *Rivista Italiana di Paleontologia e Stratigrafia*, 92, 445–478.

Garzanti, E., Nicora, A., & Tintori, A. (1986b). Permo-Carboniferous stratigraphy in SE Zanskar and NW Lahul (NW Himalaya, India). *Eclogae Geologicae Helvetiae*, 79(2), 443–478.

Garzanti, E., Haas, R., & Jadoul, F. (1989). Ironstones in the Mesozoic passive margin sequence of the Tethys Himalaya (Zanskar, Northern India): Sedimentology and metamorphism. *Geological Society, London, Special Publications*, 46, 229–244.

Garzanti, E., Angiolini, L., & Sciunnach, D. (1996). The Permian Kuling Group (Spiti, Lahaul and Zanskar; NW Himalaya): Sedimentary evolution during rift/drift transition and initial opening of Neo-Tethys. *Rivista Italiana di Paleontologia e Stratigrafia*, 102, 175–200.

Herren, E. (1987). Zanskar shear zone: Northeast-southwest extension within the Higher Himalayas (Ladakh, India). *Geology*, 15(5), 409–413.

Hutton, J. (1788). Theory of the Earth; or an investigation of the laws observable in the composition, dissolution, and restoration of land upon the Globe. *Transactions of the Royal Society of Edinburgh*, 1(Part 2), 209–304.

Marshak, S. (2019). *Earth: Portrait of a Planet* (6th ed.). New York: W.W. Norton & Company.

Myrow, P. M., Hughes, N. C., & Parcha, S. K. (2023). Revised lithostratigraphic classification of the Cambrian sequence of the Zanskar Valley, Ladakh Himalaya, India. *Journal of the Palaeontological Society of India*, 68(1), 45–62.

Nanda, M. M., & Singh, M. P. (1976). Stratigraphy and sedimentation of the Zanskar area, Ladakh and adjoining parts of the Lahaul region of Himachal Pradesh. *Himalayan Geology*, 6, 367–388.

Parcha, S. K., & Singh, B. P. (2023). Revised lithostratigraphic classification of the Cambrian sequence of the Zanskar region, Ladakh Himalaya, India. *Himalayan Geology*, 44(2), 1–18. https://journals.sagepub.com/doi/abs/10.1177/05529360231182287.

Searle, M. P. (1983). Stratigraphy, structure and evolution of the Tibetan–Tethys zone in Zanskar and the Indus suture zone in the Ladakh Himalaya. *Transactions of the Royal Society of Edinburgh: Earth Sciences*, 73, 205–219. https://doi.org/10.1017/S0263593300009688.

Searle, M. P. (1986). Structural evolution and sequence of thrusting in the High Himalayan, Tibetan-Tethys and Indus suture zones of Zanskar and Ladakh, western Himalaya. *Journal of Structural Geology*, 8(8), 923–936.

Searle, M. P., Cooper, D. J. W., Rex, A. J., & Fryer, B. J. (1988). Stratigraphy, structure and evolution of the Tibetan-Tethys zone in Zanskar and the Indus suture zone in the Ladakh Himalaya. *Transactions of the Royal Society of Edinburgh: Earth Sciences*, 79(2–3), 67–88.

Shellnutt, J. G., et al. (2011). No link between the Panjal Traps (Kashmir) and the Late Permian mass extinctions. *Geophysical Research Letters*, 38 (L19308).

Shellnutt, J. G., et al. (2023). Geochronology and geochemistry of the Panjal Traps from the southern Pir Panjal Range, Kashmir, India. *Lithos*, 436, 106967.

Smith, W. (1815). A delineation of the strata of England and Wales, with part of Scotland...[Map]. John Cary.

Thakkar, M. G., Chauhan, G., Padder, A. H., Parcha, S. K., et al. (2023). Geoheritage merits of the Zanskar range of the Kashmir Himalaya: A field geology museum from precambrian to present. *Geoheritage*, 15(3), Article 75.

Treloar, P. J., & Searle, M. P. (Eds.). (2019). *Himalayan Tectonics: A Modern Synthesis*. Geological Society, London, Special Publications, 483, 215–254. https://doi.org/10.1144/SP483.7.

Winchester, S. (2001). *The Map that Changed the World: William Smith and the Birth of Modern Geology*. HarperCollins.

Geology and Tectonics of the Kohistan-Ladakh Island-Arc-Terrane, and the Southernmost Karakoram Terrane

5

This chapter focuses on two terranes within the tectonic jigsaw collage of geological terranes, once widely dispersed, now brought together through the Himalayan orogeny. The reader is taken step by step through the key geological units that comprise the Kohistan-Ladakh Island Arc and the southern part of the Karakoram terrane.

KOHISTAN-LADAKH ISLAND ARC TERRANE

The Jurassic-Cretaceous to Oligocene Kohistan-Ladakh Island Arc terrane stretches along strike across the Trans-Himalayan mountains of Afghanistan, Pakistan, and India. The Kohistan-Ladakh terrane (KLA) (Figure 5.1) is bounded by two sutures (faults marking the junction between tectonic plates), named the Indus and Shyok Sutures. Key overview synthetic references for Kohistan-Ladakh include Petterson, 2010, 2019, and 2023 and Treloar and Searle, 2019, together with the numerous, comprehensive references cited therein. This chapter predominantly utilises these synthetic references for contextual literature-supported data, but primarily focuses on utilising field geology to provide readers with 'direct' experiences, seeing through the eyes of a geologist. For more generic geological scientific discussions we refer readers to appropriate chapters of Marshak (2019), such as chapters 4 and 6 (other similar textbooks are available). This chapter takes the reader on a traverse across the KLA, beginning with rock units associated with the Indus Suture, and progressing to the Indus Molasse and the Kohistan Batholith, then crossing the Shyok Suture to the southern part of the Karakoram Plate/Terrane.

Sutures mark the closure/subduction of oceans and collisions between continental or arc terranes. Suture zones contain obducted relics of subducted ocean plate geology (mantle ultramafic igneous rocks, ocean-floor volcanic basalts, deep-sea sedimentary rocks, and deformed components of the colliding continental/arc terranes, Marshak, 2019). Rock units spatially associated with the Indus Suture include: (1) the Lamayuru Complex; (2) the Dras Volcanics; and (3) Melange and ophiolite lithologies (Figure 5.1).

The Lamayuru Complex is Triassic-Cretaceous in age, comprising 1,000–1,500 m thick sequences of metamorphosed shales, siltstones, sandstones, and limestones. Much of the sequence has a *turbidite*

DOI: 10.1201/9781003794868-5

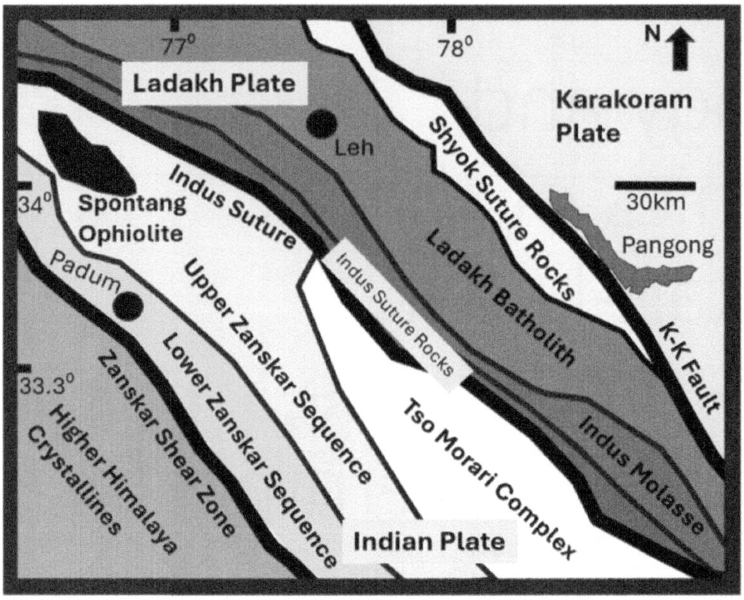

FIGURE 5.1 Geology of Ladakh: for this chapter, note the location and structural-stratigraphic position of the Indus Suture Rocks, the Indus Molasse, Ladakh Batholith, Shyok Suture Rocks, and the Karakoram Plate.

character (sediments laid down in moderate to deep ocean depths by sediment-charged, ground-hugging water currents, Marshak, 2019). Dras Volcanics comprise Jurassic-Cretaceous volcanic rocks which span a compositional range from basalt to andesite and rhyolite, typical of island arc volcanic sequences (Petterson, 2019). Melange and ophiolitic rocks are common in suture zones. *Melange* literally means a 'tectonically mixed up lithology', comprising mudstones, breccias, and conglomerates, often highly deformed and sheared. Mixed larger clast populations may include limestone, volcanic rocks, and ocean floor/mantle materials. Melanges are deposits formed during the final collisional stages of terranes. *Ophiolites* (tectonically emplaced former-oceanic-floor geology, now located within continents) are ocean floor sequences mainly comprising basalt-volcanic and mantle ultramafic-plutonic rocks, and deep-sea siliceous sediments (cherts). Melanges and ophiolites tend to be 'patchy' and discontinuous along strike. The Indus Suture has many examples of ophiolite and melange lithologies in Afghanistan, Pakistan, and India.

Figure 5.2, lower photograph (view looking northwards), takes us from the carbonate-sandstone Upper Zanskar Sequence in the south to the darker-coloured Lamayuru Complex and Dras Volcanics, to the north (Zanskar Gorge area). The upper photograph illustrates an exposure of vertical, highly cleaved Lamayuru Complex strata. At this locality, Lamayuru Complex rocks have a shiny-sheen appearance, reflecting the growth of mica and related layer-silicate minerals within low-grade metamorphic phyllites and meta-pelites/psammites. The rocks are thinly bedded, with a strong penetrative cleavage. The vertical-sub-vertical dip of the rocks increases the risk of rock/landslides adjacent to the road. Abundant layer-silicate minerals within the slates lubricate potential slipping planes, further increasing the risk of sliding, as does the regular freeze-thaw weathering climate, and snow-melt during the summer. Road engineers and geologists map lithologies and structural geology as road construction progresses, with the objective of locating potential hazards, optimising routes, and informing required engineering styles. Differing lithologies and geological structures require bespoke engineering solutions. For example, blasting approaches, slope-engineering, and selection of optimal road trajectories will vary according to geological and topographic conditions. At this locality, within a gorge-narrow river valley setting, and limited flat land availability, choices for road routes are highly constrained. Structures such as faults, fold axial planes, shear zones, and fractures are treated with engineering caution. Additional drainage and slope-stability engineering/strengthening may be required to mitigate landslide risk.

FIGURE 5.2 Lower photograph (looking northwards): well-bedded stripy light-coloured Upper Zanskar Sequence in the foreground, with overlying darker-coloured Lamayuru Complex turbiditic mudstones in the distance. Upper photograph of cleaved phyllitic slates and meta-siltstones from the Lamayuru Complex. Zanskar Gorge. *Source*: Author's photographs.

FIGURE 5.3 The Lamayuru Gompa (Monastery) is built upon and around a prominent exposure of Lamayuru Complex Schists. Lamayuru. *Source*: Author's photograph.

The Lamayuru Complex, now mostly preserved within complexly deformed structural-tectonic slices, records Triassic-Jurassic Indian passive margin geological sequences and palaeo-environments, including deep-marine turbiditic mudstones, siltstones, sandstones, and carbonates. Youngest sequences record shallow carbonate shelf environments which became progressively unstable, creating conditions for conglomerate deposition (Robertson and Sharp, 1998, and Searle et al., 1988).

Figure 5.3 illustrates a large-scale outcrop at its name-location. The Lamayuru Gompa (a Buddhist monastery and temple complex) wraps around and surmounts the Lamayuru Complex strata. The silvery-white, brown, and grey colours indicate thin-medium scale bedding and metamorphic foliation structures. Rocks are folded at a range of scales and are strongly cleaved. Differential weathering on steep slopes picks out partings produced by cleavage within the rock mass. Natural drainage channels pick their routes via cascading cleavage planar structures. This mountain-scale outcrop presents two well-developed penetrative planar structures within the same rock mass. Foliation structures, as described above, probably reflective of primary bedding, are picked out by the stripy darker and lighter coloured layers, dipping from the upper right of the cliff outcrops to the lower left in the image. Cleavage planes are particularly clearly exposed beneath the uppermost temple buildings, on the right-hand side of Figure 5.3. Cleavage planes dip sub-vertically towards the observer, forming almost vertical planar cliff faces.

Figure 5.4 presents a range of characteristics of the Lamayuru Complex lithologies close to Lamayuru village. The left-hand-side photograph displays the macro-character of the outcrop. Sub-vertically dipping, thinly bedded/foliated, cream-light-grey-coloured siltstones and sandstones crop out adjacent to the road. Middle and right-hand side photographs are close-ups of nearby outcrops. Here, the sandstones are red-coloured siltstones, interbedded with thinly laminated siliceous sandstones. Quartz and silica veins are abundant. Small-scale folds display disharmonic geometries. Axial plane directions change along the length of the folds. The middle photograph illustrates a 'Z' shaped fold produced by top-to-right lateral simple shear along the limb of a larger fold. Geologists use structures such as this to predict the existence of larger-scale folds. In this example, an antiform may exist to the upper right and a synform to the lower left. The left-hand side of the fold has been sheared (adjacent to the pencil in the picture): this c. 5 mm wide zone has highly stretched the rock mass and broken up the continuity of the fold structures. A second shear zone is exposed to the right of the 'Z' shaped fold structure. The right-hand side photograph presents an 'M' shaped antiform with tight-isoclinal limbs. 'M' and 'W' shaped folds are typical of fold hinge zones and fold closures. 'Z' and 'S' shaped folds occur on the limbs in between the 'M' and 'W' shaped folds. These structures allow geologists to unravel complex architectures of folded strata: the use of minor structures allows geologists to map larger-scale structures.

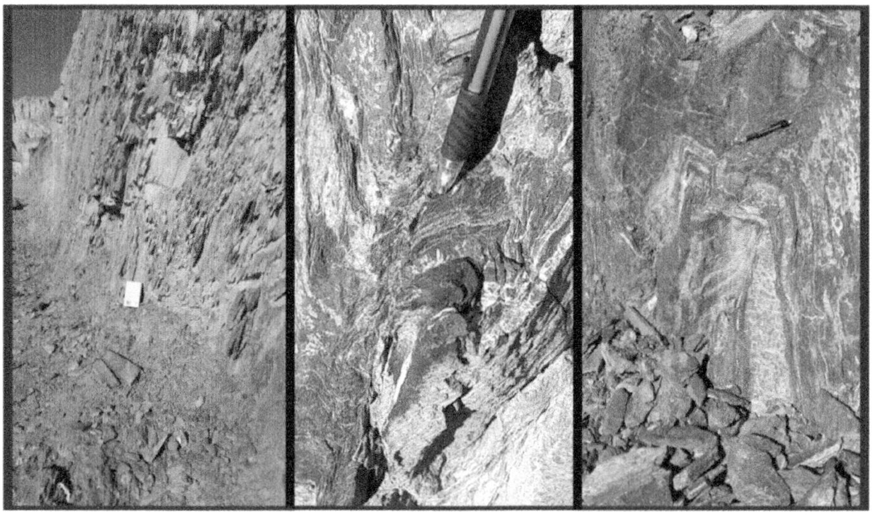

FIGURE 5.4 Thinly bedded/foliated Lamayuru Complex siltstones, mudstones and sandstones, folded at a range of scales. Lamayuru. *Source*: Author's photographs. The notebook in the left hand side photograph measures 19 x 12cm. The pen exposed in the central image imeasures 10cm. The pencil in the right hand side image measures 15cm.

The Dras Volcanics are the first geological unit encountered thus far which are non-Indian Plate unit: they are an integral part of the KLA, and largely crop out close to the Indus Suture within Ladakh. Dras Volcanics are possible correlatives with the Chalt Volcanics in Kohistan: both units are aged Jurassic-Cretaceous. Dras Volcanics are comprised of a wide range of lithologies, from voluminous basalts to andesites, dacites, rhyolites, and pyroclastic rocks, with volcaniclastic sandstones, siltstones, and vitric tuffs. Alongside field-outcrop descriptions of Ladakh Dras Volcanic exposures, this chapter explores related genetic processes and tectonic environments. For further details of the Dras Volcanics, readers are directed to Petterson, 2019, Robertson and Sharp, 1998, and Walsh et al., 2019.

Figure 5.5 presents three main lava units of the Dras Volcanics, each around 4–6 m thick. Every unit marks a separate lava flow event. Continuous white lines mark a part of the contact zone between lava flows, the middle flow highlighted, lying between upper and lower volcanic units. A lack of other deposits between lava flows indicates little or no time gaps between individual lava flow (effusive) events, also suggesting rapid eruption/effusion activity. The middle lava displays 'flow joints', some of which are marked by pecked white lines. Flow joints indicate the direction of lava flow: in this case from left to right in Figure 5.5. As lavas cool, they become more viscous, with a more sluggish flow forming and 'freezing' flow joints. The outer carapace of lava, in contact with the atmosphere or the substrate over which the lava flows, cools, and solidifies most quickly, forming solid crusts of variable thickness, whilst the lava interior zone remains molten. These skins or crusts often break up through the action of the molten interior, tearing the crusts apart and *brecciating* to create *flow breccias*. The uppermost unit in Figure 5.5 is brecciated in parts.

Dras Volcanics crop out adjacent to the Indus Suture Zone. Widest outcrops occur in the west, close to Dras. Volcanic outcrops thin eastwards. The volcanic lithological character changes from west to east. Western areas are dominated by *lava flows,* whilst eastern outcrops are dominated by *volcaniclastic deposits*. Lava flows result from the eruption of *magma* (largely liquefied rock with dissolved volatiles) at the surface of the Earth via point source(s) (e.g., stratovolcano or fissure [volcanic cracks] effusive eruptions). Lava flows follow gravity, flowing down volcano flanks and surrounds. The shape and form of lava flows depend on a wide number of variables, including the composition of lavas, the viscosity of lavas, the volatile

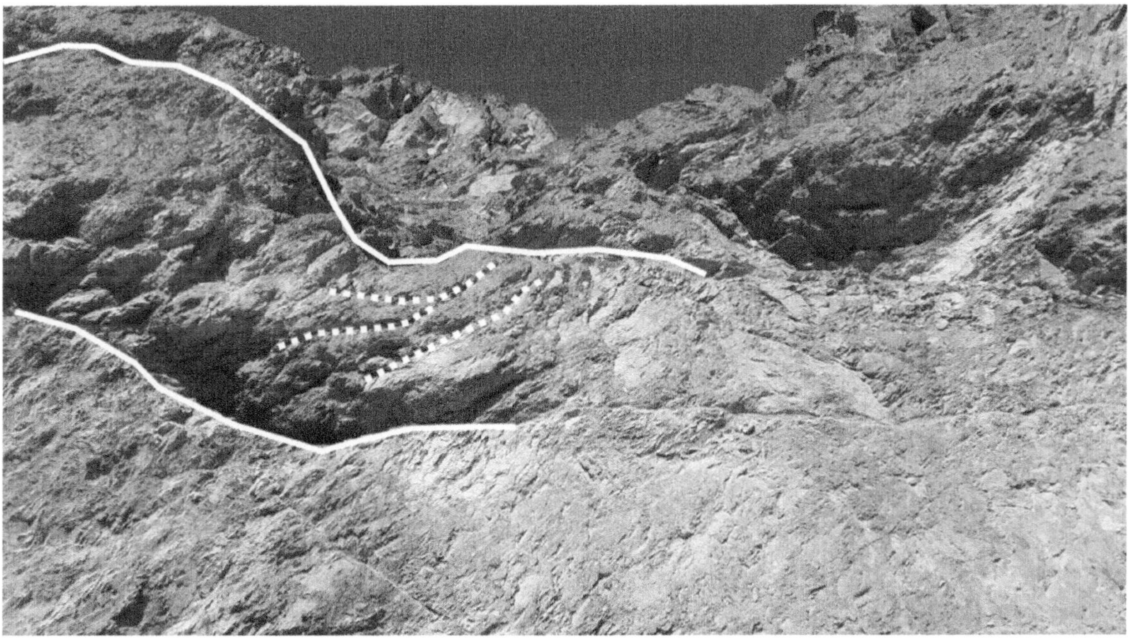

FIGURE 5.5 Successive Dras volcanic lava flows. *Source*: Author's materials. Note: White continuous lines mark part of the contacts between lavas. White pecked lines mark flow joints within the central lava. Pashkyum-Mulbeck Road. Central flow is around 5m thick.

content of lavas, and the volume erupted at any one time. In general, high-temperature, low-viscosity lavas flow most easily and at highest velocities, and can act somewhat like water and rivers, even locally displaying 'waterfall' type characteristics. These types of lava can flow for long distances and be relatively thin. Accumulation of fluid basaltic lavas produces low-angle shield-shaped volcanoes. More viscous lavas tend to flow slowly, at lower temperatures, producing thick semi-solid crusts and forming relatively thick lava piles. These more viscous andesite-rhyolite lavas produce steep, upstanding, conical-like *stratovolcanoes*. Very high-volume eruptions of low-viscosity lavas overwhelm and bury landscapes over which they flow: rapid, regular eruptions produce thick volcanic plateaux, forming 'trap' topography, such as the Deccan Traps in peninsular India. *Volcaniclastic* rocks are those that have a volcanic origin but are comprised of particles/clasts rather than crystals. Volcaniclastic rocks result from more violent, gas-charged, explosive eruptions: the most dramatic of which produce highly dangerous, low-viscosity, extremely fast-moving, ground-hugging, fluidised, particulate *pyroclastic flows*. Other deposits result from red-hot solidified rock fragments violently blasted into the air from volcanic vents. On exit from the volcano, hot *pyroclasts* (hot volcanic particles/clasts) follow parabolic travel trajectories under the influence of gravity, eventually settling on land or subaqueously. These formerly airborne solid eruptive particles/clasts form layered deposits which drape (and eventually bury) topography: termed *pyroclastic fall* deposits. Volcanic clasts/particles can be reworked through the action of rain, rivers, air, glaciers, and marine processes, producing turbidites, conglomerates, sandstones, siltstones, and mudstones. Volcanic rocks that display reworked sedimentary facies (but are comprised largely of original primary volcanic material) are termed *secondary volcaniclastic or epiclastic volcanic rocks*. See Marshak, 2019, chapter 5, for further details.

The Dras Volcanics have possible correlative units in Kohistan (Pakistan): they are named the *Chalt Volcanics*. The Dras (and Chalt) Volcanics are examples of *island arc volcanism*. *Island arcs* form when one oceanic plate subducts beneath another oceanic plate, often at some distance from any continent, generating melting and magmatism. These magmas accumulate and construct islands and submarine edifices, which, from a bird's-eye perspective, form a series of islands/edifices within an arc-like geometry: hence the term *island arcs*. Island arcs are one of the main tectonic mechanisms for the production and growth of new continental crust. As arcs grow and thicken, they become relatively buoyant, creating longer-lasting continental crust. Island arcs ultimately collide and suture with larger continents, as has happened in Ladakh. The relative buoyancy of arcs reflects the predominant andesite-granodiorite composition of volcanic-plutonic igneous rocks that form the bulk of island arcs. To understand how island arcs become buoyant, it is instructive to understand the basics of igneous rock classification (see Marshak, 2019, chapter 4 for more details). Igneous rocks are the solidified product of *magmas* (liquid rock). Most magma is produced by the melting of the Earth's mantle, the layer beneath the Earth's crust, forming over 80% of the interior of the Earth. Once this primary magma (termed basalt) is produced, it can evolve and change through a range of igneous processes, such as *fractional crystallisation*. The simplest way of classifying igneous rocks is through their silica (SiO_2) content, their major mineral composition, their colour, and whether or not magmas reach the Earth's surface. Magmas that reach the Earth's surface are termed *volcanic*, whilst rocks that freeze at depth are termed *plutonic*. For this book, we will restrict ourselves to eight terms that describe igneous rocks (volcanic/plutonic): *basalt/gabbro*; *andesite/diorite*; *dacite/granodiorite*; and *rhyolite/granite*. These terms distinguish finer-grained volcanic rocks from coarser-grained plutonic rocks. For example, a *rhyolite* is a volcanic rock, whilst a *granite* is a plutonic rock. In terms of silica (SiO_2), basalts and gabbros have SiO_2 compositions of between 45% and 54%, *andesites* and *diorites* have SiO_2 compositions of between 54% and 66%, *dacites* and *granodiorites* have SiO_2 compositions of between 64% and 73%, and *rhyolites/granites* have SiO_2 compositions of >73%. In terms of mineralogy, basalts and gabbros contain a relatively high abundance of darker minerals richer in iron and magnesium, such as *olivine, pyroxene,* and *amphibole*, whilst rhyolites and granites are dominated by lighter-coloured minerals rich in SiO_2 and aluminium, such as *feldspar* and *quartz*. Andesites/diorites occupy the middle spectrum between basalts/gabbros and rhyolites/granites with respect to the darker/lighter coloured mineral ratio. Gabbros and basalts are relatively darker in colour and relatively dense because of the higher abundance of iron and magnesium-rich minerals. Rhyolites and granites are relatively lighter in colour and less dense, as they contain a higher proportion of aluminium and SiO_2-rich minerals, with lower specific gravities.

As island arcs evolve and thicken, they become increasingly composed of rocks from andesitic/dioritic to rhyolitic/granitic compositions. Their relative density falls, and buoyancy increases. From the middle of oceans, island arcs and new continental crust are thus created. The great bulk of island arc crust is not subducted back into the mantle and ultimately becomes a permanent component of continental crust, as is the case with the KLA, now an island arc terrane within the Eurasian-African modern-day supercontinent.

Dras Volcanics comprise rock compositions from basalt to rhyolite, although basalts and andesites dominate. The petrological-chemical composition of the Dras Volcanics is typical of island arcs. Figure 5.6 presents three thick andesitic lava flows, with the central lava being around 8–10 m thick. The central lava displays strong vertical joints typical of lavas that cool from a myriad of cooling cells from the top to bottom of a lava flow, producing hexagonal columns throughout the lava. These columns are separated by vertical joints (termed *columnar jointing*) as in Figure 5.6. Continuous white lines mark the contacts between lava flows. A lack of intervening deposits between lava flows indicates rapid effusion rates. A series of stepped lavas, such as Figure 5.64, form stepped flat-lying or gently dipping topography termed *trap topography*.

Figure 5.7 presents *pillow lava* structures. The upper left photograph is an enlargement of the right-hand side photograph. The term *pillow* refers to the hemispherical structures within the lavas presented in Figure 5.7. Whilst undeformed pillows come in a range of shapes and sizes, the 'ideal' is a bulbous-hemispherical shape in three dimensions. The origin of the term *pillow lava structures* goes back in the literature to the early 20th century and even to an 18th-century account of the shape of lavas around Mount Vesuvius in Italy (Hamilton, 1772). Whilst the formation of the structures had not been directly witnessed, geologists correctly interpreted the structures as indicating a submarine environment of eruption. Underwater divers in the 1970s filmed pillows forming from lava within an underwater setting (Moore, 1975). As lava erupts under water, it experiences rapid cooling and forms thin glassy crusts/carapaces. Water has a heat specific capacity 4.2 times greater than air. Some cooling systems using water claim they can be up to 3,000 times more efficient than air. As magma extrudes on the bottom of the ocean, it is exposed to enormous volumes of cold water which rapidly absorbs and cools volcanic heat. Once the glassy crust forms, it can be inflated in a similar way to blowing up a balloon. The surface of submarine lavas cracks, and deeper molten lava breaks through the cracked surface, forming pillows. As pillows grow, they too can crack and 'bud' other pillow structures. As well as being used as a palaeo-environmental indicator, pillows allow geologists to work out which 'way-up' strata are. Pillows form curved convex-upwards upper surfaces that indicate the original 'right-way-up' of the volcanic stratigraphy. The upper left- and right-hand photographs, for example, are the right way up, as the pillow structures have

FIGURE 5.6 An andesitic lava pile is present within the Dras Volcanics. *Source*: Author's materials. Note: The central lava displays columnar jointing, particularly well illustrated on the left-hand side. The outcrop presents stepped volcanic plateau-like topography termed trap topography. Pashkyum-Mulbeck Road. Central lava is 8-10m thick.

FIGURE 5.7 Pillow lavas within the Dras Volcanics. Pashkyum-Mulbeck Road. *Source*: Author's photographs and interpretations. 25cm of a metre ruler are exposed in the upper left hand side and right hand side photograph. 35cm of the metre ruler is exposed in the lower left photograph.

upward-facing convex upper surfaces. The lower left photograph presents deformed pillows. Here, pillows have length/width ratios of 4 to 5:1, whilst the less- or undeformed pillows in the upper left- and right-hand side photographs have length/width ratios of around 1.3. Pillow structures are used by geologists as strain indicators and quantitative data are incorporated within methods of measuring strain experienced by the rocks during deformation events. In this case, for pillow length/width ratios to change from 1.3 to 4–5:1, the rocks must have experienced c. 300-400% strain or change of shape from the original to the final shape in percentage terms. This example gives a palpable impression of the extreme deformations involved during orogenesis.

Figures 5.8 and 5.9 exhibit Dras Volcanics volcaniclastic *facies* (facies is a term that refers to the environmental context of a deposit, e.g., a deposit formed within a desert or shallow marine facies/setting). Figure 5.8 A presents a gritty, poorly sorted sandstone facies. Figure 5.8 D illustrates thin to moderately thin bedded (from mms to 10 cm) volcaniclastic sandstones-siltstones, from turbidite sequences. Figure 5.8 B exhibits dark, lozenge-shaped mud flakes with aligned long axes and a background grey siltstone, deposited within a marine or freshwater turbidite unit. Figure 5.8 C presents ripple structures: hummocky curvilinear beds, particularly visible in the lighter coloured beds, indicating a shallow submarine facies shaped by water currents.

Figure 5.9 illustrates thin-bedded glassy-vitreous, parallel-bedded, fine-grained, volcaniclastic rocks. Differential, alternating pale- and darker-coloured beds allow for clear distinctions in bedding to be observed. This sequence can be interpreted as distal (remote) airfall volcanic ash, which has been *well-sorted* (clasts do not vary much in size) as it travelled through the air from the original volcanic eruptive source, with only the finer particles (often of a vitreous composition) travelling long distances from the volcano. Coarser-clast fractions would have dropped out from the airborne volcanic ash-cloud, at proximal localities (closer to the volcanic source). Additional settling through a column of water can enhance sediment particle sorting. Alternatively, the sequence may be a distal turbiditic volcaniclastic sediment.

There are numerous examples of turbidites within Zanskar and Ladakh. Turbidites occur within the High Himalaya Crystallines, the Zanskar Sequence, the Lamayuru Complex, and the Dras Volcanics. Turbidites are common sedimentary rocks (Marshak, 2019, chapter 7). They are characterised by graded

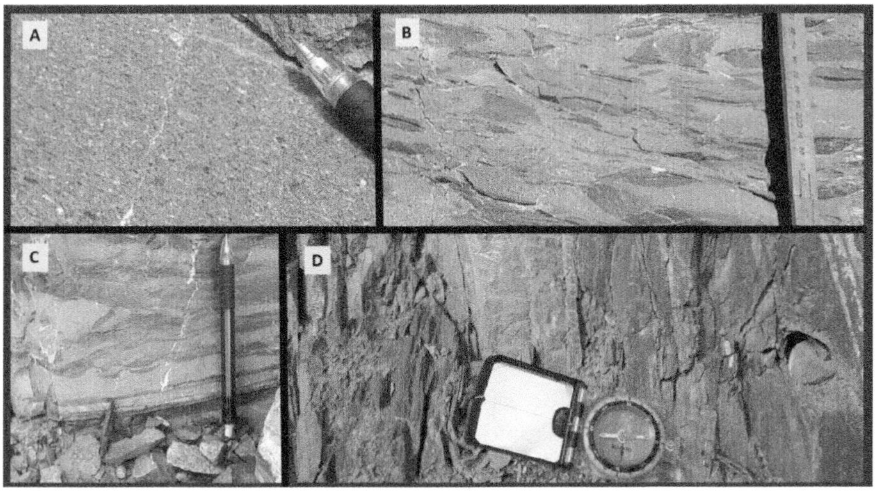

FIGURE 5.8 Different volcaniclastic facies within the Dras Volcanics. *Source*: Author's photographs. Note: Lamayuru to Khalsi Road. See text for details. Photograph A presents 3cm of a pen for scale. Photograph B presents 16 cm of a metre ruler. The pen in photograph C is 15cm long. The extended compass in D is 20cm long.

FIGURE 5.9 Siliceous bedded volcaniclastic Dras volcanics. *Source*: Author's photograph. Note: Lamayuru to Khalsi Road. See text for details. The pen is 15cm long.

bedding sequences that may vary between coarse conglomerates at the base and fine mudrocks at the top. Grading here means a change in grain size with sedimentary height: as we observe from the bottom to the top of a rock sequence, we see a perceptible change in grain size. This change can be coarse to fine, or fine to coarse, depending on the local sedimentation conditions. Turbidites are produced from sediment-charged turbid-water currents. Turbidite currents are opaque to translucent, as they contain a high proportion of sediment. Turbidites move, through the force of gravity, from higher submarine/subaqueous levels to deeper environments. They may form fans underwater (lake, sea, or ocean) or move ever deeper into distal (remote) ocean plains or down long submarine canyons. As the momentum of the current diminishes, sediment deposits at the base of the water current. Coarser, less sorted sediments such as conglomerates are deposited first. These deposits grade upwards through coarse-fine sandstone and siltstone. Mud can settle on the top of the turbidite sequence if there is a sufficient time gap between one turbidite event and a subsequent one. Turbidites that are proximal (close to) to the original sediment source of the turbidite current are thicker and coarser grained. Distal turbidites are more thinly bedded and finer grained. As one turbidite impinges on an earlier turbidite deposit, it may erode the top of the underlying deposit and can, for example, break up an upper mud layer. Fragments of the eroded and disrupted lower mud layer are then incorporated into the sand-rich body of the overlying turbidite. Turbidite currents are highly energised and regularly break underwater cables. Deep-sea cabling companies do their best to lay cables in areas of low turbidite activity or bury the cable to a depth where they are less likely to be impacted by turbidity currents.

Evidence presented here for the Dras Volcanics indicates a range of palaeo-eruptive environments and palaeo-eruptive styles. The bulk of the Dras Volcanics comprises a pile of lavas that vary in composition from basalt through to rhyolite. Some of the lavas were erupted sub-aqueously. Others may have erupted above water. Thick lavas with no intervening material between lavas indicate periods of high effusion rate. Volcaniclastic facies indicate pyroclastic activity alongside effusive volcanic activity. The bulk of the volcaniclastic material was 'reworked': i.e., once generated by primary volcanism, it became remobilised by the action of water and wind and redeposited, perhaps in moderate to deeper ocean depths via turbidity currents.

Melange units' crop out at numerous intermittent localities and within more contiguous outcrops along the Indus Suture Zone. Figure 5.10 presents a sheared, *heterolithic* (clasts of varying lithologies) conglomerate. The term *heterolithic* applies to the varied nature of large-clast lithologies. Rock types such as granite, diorite, chert, sandstone, basalt, and ultrabasic rocks make up the varied assemblage of the larger-clast population. Larger clasts vary in size, up to c. 30–40 cm, with most clasts being between 3 and 6 cm. Clast shape varies between rounded and subrounded to subangular. The matrix comprises highly sheared mudstone and siltstone. A particularly strong shear zone is present in the central part of the figure, dipping from upper left to lower right. A close inspection of the whole photograph in Figure 5.10 reveals the highly sheared nature of the matrix, with the presence of numerous fractures and stretched materials. Shear fabrics wrap around some of the larger clasts.

Figure 5.11 illustrates examples of close-ups of the larger clast lithologies. The right-hand side photograph illustrates an angular blue-black basaltic clast with smaller dark angular larger clasts and sheared/folded matrix with quartz veins. The left-hand side photograph presents more rounded clasts and variable lithologies. Melanges form within collisional zones and help geologists to map and define suture zones (Marshak, 2019, chapters 4 and 11). As larger fault blocks and terranes collide, intervening material is ground up, crushed, and sheared. A variety of materials will be present within these highly dynamic tectonic environments. A range of crustal depths will be exposed, facilitating the erosion of shallow and deep crust, alongside obducted oceanic and mantle materials. This environment hence produces sheared lithological mixtures of widely available geological materials.

FIGURE 5.10 Highly sheared melange. *Source*: Author's photograph. Note: The range in clast size within the melange (poorly sorted) and the sheared nature of the matrix. Khalsi area. 25cm of a metre ruler for scale is located within the image.

FIGURE 5.11 Heterolithic nature of Indus Suture Melange. *Source*: Author's photographs. Note: Large-clast lithologies include granite, sandstone, chert, diorite, basalt, and ultramafic rocks set within a sheared mud and siltstone matrix. Khalsi area. The compass in the left hand side image measures 20cm. 9 x 6cm of a field notebook is exposed in the right hand side image.

FIGURE 5.12 Indus Suture Zone asymmetrical folding. Nimu. Source: Author's photograph.

Suture zones are areas of intense deformation alongside lower levels of strain. Tectonic plates collide at mm to cm per year, producing massive accumulations of strain over millions of years within the collisional (suture) zones along plate boundaries. This accumulation of tectonic strain leads to brittle faulting and fracturing within upper crustal units and more ductile folding and shear structures at deeper levels. Figure 5.12 illustrates strongly asymmetrical folds within the Oligocene-Miocene turbiditic sandstone-siltstone Nimu Formation of the Indus Suture Rocks, close to the confluence of the Indus and Zanskar Rivers. A Recent unconformity is visible with horizontal alluvial deposits overlapping the fold structures. Asymmetrical folds provide indicators of local tectonic-stress vectors at the time of deformation. They give a sense of *vergence:* the overturned nature of this fold geometry indicates the longer upper limb pushed and stretched (sheared) according to tectonic stress vectors, in this case from left to right in Figure 5.12.

Figure 5.13 illustrates a mountain-scale overturned tight to near isoclinal fold: almost a nappe structure. The fold verges from right to left, towards the north, with a highly thinned and sheared lower limb, adjacent to a several km-scale shear zone. This structure records local tectonic thrust, shear, and fold vergence sense at the time of Indus Suture formation, and during the later stages of India-Ladakh plate collision.

Figure 5.14 is from the Khalsi area, presenting a cross-section of a nappe structure within the Indus Suture Rocks. Small-scale folds are visible on the upper limb of the nappe. The geometry and orientation of the nappe indicate motion towards the east-southeast, and local strike-slip motion within the Indus Suture. Sutures form over a significant period of time within moving stress fields. Whilst a dominant tectonic stress field was north-south, local tectonic movements can deviate from the predominant stress field.

Whilst it can be challenging to precisely locate the Indus Suture within Ladakh, the position of the suture is clearly visible in parts of the Kohistan mountains, Pakistan. Figure 5.15 is one example of a

FIGURE 5.13 Mountain-scale recumbent folds close to the Indus Suture Zone, Zanskar Valley. *Source*: Author's photograph and interpretation.

FIGURE 5.14 Roadside recumbent-nappe fold structure within the Indus Suture Zone, Khalsi. *Source*: Author's photograph and interpretation.

FIGURE 5.15 Exposed Indus Suture, Karakoram Highway, Kohistan, North Pakistan. *Source*: Author's photograph and interpretation.

FIGURE 5.16 View from Lamayuru looking southwards over steeply dipping Indus Suture Zone Rocks, towards the Indian Plate Zanskar Sequence in the distance. *Source*: Author's photograph.

well-exposed section through the actual Indus Suture. Here, the Indian Plate rocks are directly in contact with the Kohistan Island Arc rocks, which comprise the Jijal Complex (an obducted part of the upper mantle) and the Kamila Amphibolites (meta-volcanic-sedimentary rocks). In Figure 5.15, the Kohistan Plate is thrust southwards over the Indian Plate, which deforms as it 'subducts' beneath Kohistan: an isoclinal, downward-facing nappe fold structure is visible within the lower Indian Plate, indicating motion downwards towards the north. It is indeed awe-inspiring, alongside a humbling experience, to be able to actually touch a plane where two great tectonic plates meet!

This section ends with Figure 5.16, a photograph looking southwards from Lamayuru towards Zanskar. The foreground comprises steeply dipping Indus Suture Rocks. The steep dips to the south reflect thrusting towards the north. Steeply dipping, densely thrusted and sheared geological sequences typify suture zones. In the distance, Indian Plate sedimentary rocks of the Zanskar Sequence are exposed in the higher mountains. As this section ends, the narrative now moves northwards through the Ladakh Island Arc.

LITHOLOGIES, FIELD OUTCROPS, AND LANDSCAPES OF INDUS MOLASSE ROCKS

Figure 5.1 presents the location of the Indus Molasse Rocks: coloured in brown, with the Indus Suture Rocks to the south and the Kohistan batholith to the north. The outcrop width of the Indus Molasse

Rocks varies between 15 and 20 km and extends NW-SE for over 150 km. For much of its outcrop, the Indus Molasse forms mountains to the south of the River Indus. Figure 5.17 presents a simple cartoon of the formation of molasse sediments and rocks. Indus Molasse Rocks vary in age between Eocene and Miocene (c. 48–6 Ma). Lithologically, they comprise a complex sequence of conglomerates, breccias, coarse sandstones, siltstones and mudstones. Interpreted sedimentary facies include meandering-braided river, alluvial fan, high-energy fluvial, fan-delta, and delta turbidite environments. Readers are directed to Petterson 2023 and references therein.

As mountains rise through thrust-fault tectonics, they are eroded by glaciers and deep-cutting rivers, depositing proximal alluvial fans that form between mountain fronts and valleys. Rivers erode and form 'V' shaped valleys and ultimately discharge sediment into wide valleys and valley flats within the mountain front plain area. Sediments vary tremendously from boulders, cobbles, and gravels, to gritty coarse sands, fine-grained sands, silts, and muds. Wind plays its part in transporting and reworking deposits. Mountains can be impacted by extensional tectonics (not only compression), which facilitates the production of wide valleys and valley floors within mountains, causing subsidence to occur, accommodating kilometres of sediment thickness in front of thrust mountains. The weight of thrust slabs and mountains depresses the crust at the leading mountain front, leading to crustal subsidence, allowing sediments to gather in great thicknesses and extent within a *foreland basin*. Figure 5.17 illustrates the geomorphological setting of a foreland basin in front of actively rising mountain fronts. The Indo-Gangetic Plain of India and Pakistan is a classic example of foreland basin molasse sediments that form in front of the Himalayas. Sediment is eroded, transported, and deposited in a myriad of forms. Alluvial fans are the most proximal deposits (Figure 5.17) produced by rivers and streams that erode mountain channels and form wedge-like, thick, prismatic deposits of boulders, blocks, pebbles, gravel, sand, and mud at the foot of the mountain front. These deposits form arcuate fan-shaped features. Rivers increasingly transport and sort sediment as they move some distance from the mountains onto the valley plain. Rivers deposit broad lenticular deposits of gravels, sands, and muds that accumulate, creating flat-topped alluvial plains. Within the plains, standing bodies of water form lakes. Here, muds settle and form well-sorted, very fine-grained muds. Turbidites also form within lakes. Ice and glaciers may be significant landscape-shaping agents in the higher, colder mountains.

The Indus Molasse is 2,000–3,000 m thick in total, recording a wide range of palaeoenvironmental facies, including alluvial fan, alluvial delta, turbiditic, lacustrine (lake), and fluvial (river) facies. Figure 5.18 presents an image of a valley cut NNE-SSW into the Indus Molasse. Bedded rocks that vary in colour from red to brown to grey to white dip to the north (into the distance in the photograph) and strike right across the Stok Valley. Figure 5.18 presents how dipping rocks 'V' in the direction of dip across a valley. The valley itself has a classic 'V' shape, typical of river valleys. Significant thicknesses of Indus Molasse outcrops in Figure 5.18 are illustrated: the distant and foreground mountains are all composed of Indus

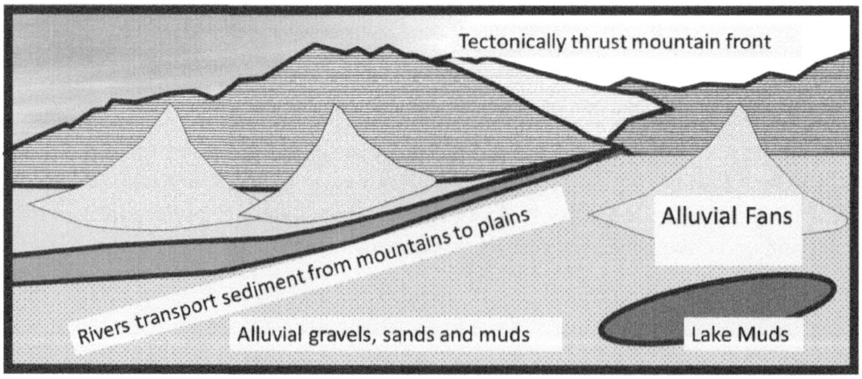

FIGURE 5.17 Cartoon pictogram presenting a simplified model for the formation of molasse deposits: as mountains rise tectonically along a thrust-mountain-front, erosion agents (ice, water and wind) erode, transport, and deposit sediment within a foreland basin at the foot of the mountains, creating intra-montane plains and thick sedimentary deposits. *Source*: Author's pictogram.

Molasse rocks. The valley cuts through tectonically uplifted and thrust mountains that have been moved northwards and tilted southwards. The Stok River cut through the rising mountains, flowing northwards into the Indus River. In turn, tributary small streams and rivers flow into the Stok River. The bedload of the River Stok is predominantly coarse, composed of boulders, blocks, gravel, and sand, illustrating current alluvial environments. The River Stok is braided (like braided hair) with a number of smaller streams threading across the valley floor.

Figure 5.19 presents a c. 800 m long continuous valley exposure of the Indus Molasse within a northeast-southwest trending valley, situated to the east of the Stok valley presented in Figure 5.18. Northeast is situated on the left-hand side of the photo-mosaic. Consistent dips of the Indus Molasse are clearly visible: rocks dip towards the right of the image. Careful inspection reveals how the amount of dip varies across the outcrop. Dips are significantly gentler at c. 20° at the right-hand side of the image, whilst the average dip is closer to 60° at the left-hand side of the image. The structure of the rocks is more complex towards the left/northeast side of the outcrop. Contorted, folded, thrusted, and faulted strata are more common in this area, as will be shown in more detail below. Whilst the bulk of the outcrop from the middle to the right-hand side of the image consistently dips at between 20° and 30°, with individual beds tracking across the mountains and intervening valleys, the outcrop is disturbed locally by thrust faults and folds. Bedding thickness varies across the outcrop, with thicker finer-grained sandstones and siltstones weathering back in a recessed manner. Thinner, more strongly cemented sandstones and conglomerates form prominent ribs/ridges.

The broad valley floor slopes steeply from SW to NE (right to left in the image). A myriad of interweaving braid channels cuts across the valley floor with ridges of sediment in between. Channels frequently change course, although there is a modern preferred series of channels. In times of flood, sheetwash can move across the valley floor, with other streams becoming active at times of high rainfall. The desert climate is apparent: only sparse grasses and shrubs can survive in this harsh environment.

Figure 5.20 presents an interpreted satellite image of the area east of Stok, part of which is presented in Figure 5.19. Even within the satellite image, the bedding planes of the Indus Molasse can be traced across ridges. Curved bedding traces mark the shapes of folds within the strata. Thick purple-red units track across

FIGURE 5.18 Thrust mountains formed of Indus Molasse Rocks. Source: Author's photograph. Note: Rocks dip southwards and have been thrust northwards. The river valley in the foreground hosts a coarse boulder-block-gravel dominated deposit. Upper Stok Valley.

FIGURE 5.19 Great thicknesses of bedded strata cropping out on one side of a NE-SW trending valley within the Indus Molasse. *Source*: Author's photo-mosaic. Note: Photo-mosaic from a valley situated to the east of Stok Valley.

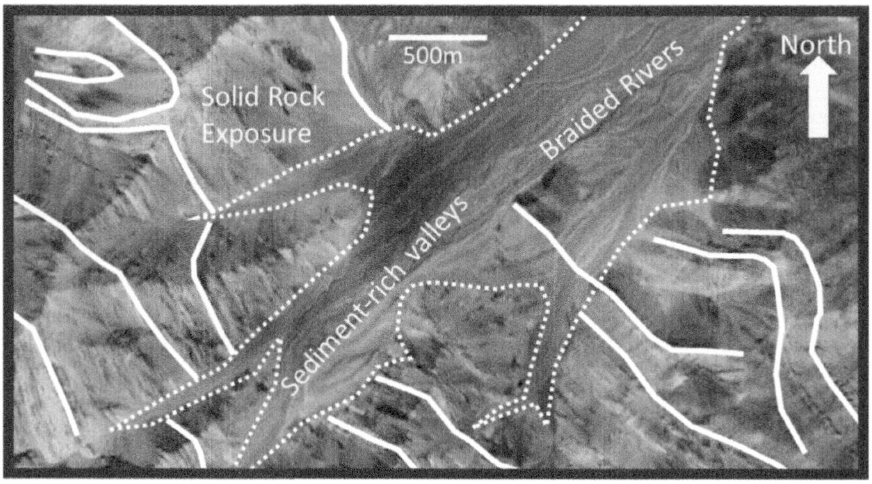

FIGURE 5.20 A satellite image of a NE-SW trending valley east of Stok. Note: White lines mark some bedding traces; others are left unmarked for readers to evaluate. Valley-superficial deposits are delineated. *Source*: Authors' materials. Base map: Acknowledgements Google Maps. Google. (n.d.). [Google Maps Stok area Ladakh, NW India. Retrieved January 10th 2025, from https://www.google.com/maps/place/Stok+94104/@34.0508013,77.525692,14381m/

the image (see the outcrop area marked 'solid rock exposure' in Figure 5.20). Pecked lines delineate the lower valley areas covered by unconsolidated gravels, boulders, blocks, gravel and sand. Numerous river channels are visible, with upstanding lenticular 'islands' of sediment between channels being clearly visible.

Figure 5.21 presents a coarse, gritty, poorly sorted Indus Molasse sandstone. The largest clasts measure around 1 cm. Finest grained clasts measure c. 1 mm and less. This variation in clast size, termed *poorly sorted,* indicates high fluvial (water) current energies and sediment-charged currents. Clast shapes vary between sub-angular and sub-rounded. Quartz, feldspar, and lithic (rock fragment) clasts are all represented within the larger clast fraction. Poor sorting and variations in clast roundness indicate proximal-medial facies for these rocks. As sediment piles thickened, the cementing action of warm silica-rich fluids, together with increasing compaction, pressure, and time, led to the *lithification* of the sediment (sediment turned into rock).

Figure 5.22 presents graded-bedded sequences of sandstone. Individual beds are 4–5 cm thick. Bedding planes are continuous across the figure. The sandstone is a poorly to moderately sorted pale white-grey unit. Largest clasts are c. 4–5 mm in diameter, with the finer, darker clasts, which clearly define bedding planes, measuring less than 1 mm. A careful observation of Figure 5.22 reveals graded bedding. Grading here is not particularly well-defined and is more like *coarse-tail* grading. This term refers to the grading of the larger clast size population, whilst smaller clast sizes do not grade. Largest clasts occur close to the base of the bed, whilst the smaller sized larger clasts occur towards the upper parts of each bed. Finer-grained clasts do not follow grading and 'dilute' the grading with larger clasts

FIGURE 5.21 Coarse, poorly sorted, gritty quartz sandstone. Indus Molasse. East Stok Valley. *Source*: Author's photograph.

throughout the rock, being surrounded by the smaller-grained silt and fine sand particles. Bedding planes are mainly horizontal and sub-parallel. Close to the ruler, towards the top of Figure 2.90, there is an upwards-convex 'hump' in the bedding. This is an example of a '*sedimentary structure*'. Sedimentary structures are produced by sedimentary processes that give clues to the palaeo-environment of deposition. For example, worms and molluscs may burrow within sediment and leave traces of their movements behind. Water currents may produce ripples within the sediment, preserved within lithified rocks.

Careful recognition, observation, and recording of sedimentary structures build up a comprehensive picture of as many elements of the depositional environment as can be gleaned from rock exposures. This upward convex sedimentary structure may be a ripple structure, indicating that a water current moved around micro-topography within the subaqueous sediment surface. Current-deposited sediment around this micro-topography deposited concave-upwards very-thin layers (termed *laminations*) of silt. The rock within Figure 5.22 is interpreted as a proximal-medial turbidite within a lake. The lake was part of a mixed environment of alluvial fans, dynamic rivers and streams, and standing bodies of water.

Figure 5.23 illustrates a conglomerate from Hemis, east of Leh. A conglomerate is a very poorly sorted sedimentary rock with a wide range in clast size. Larger clasts can include boulder, block, and pebble-sized clasts, set within a matrix of sand, gravel, and silt-grade clasts. Figure 5.23 illustrates closely to moderately packed conglomerates: large particles are positioned in reasonably close proximity to one another; sand and silt particles separate one large clast from another (matrix-supported). Most large clasts are rounded to sub-rounded with occasional sub-angular fragments. This indicates that the clasts have been moved some distance from their source and 'knocked the rough edges' off of one another. A variety of lithologies is present within the rocks, indicating that a number of geological units were being weathered and eroded in the source region. This provides clues to the source region of the conglomerates. In the case of the Indus Molasse, most clasts are from the Ladakh batholith and related volcanic rocks towards the north, indicating that rivers moving from north to south were predominant. In Figure 5.23, granite, granodiorite, micro-conglomerate, siltstone, basalt, andesite, and quartz clasts are visible; these conglomerates are classified as *heterolithic*, or *polylithic,* containing a variety of lithologies. Currents of water that deposit the protolith of conglomerates are *hyper-density* currents. That is, they are loaded with particles of all sizes. The sheer density of particles within the current means that they cannot be sorted: physically, there is 'no room' for particles to move

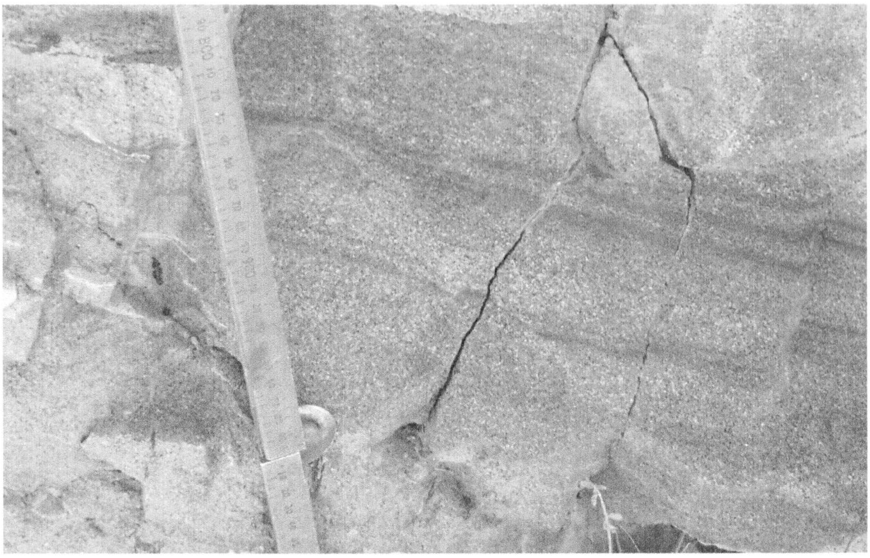

FIGURE 5.22 Graded bedded sandstone to siltstone. *Source*: Author's photograph. Note: Indus Molasse. Some beds display graded bedding, particularly within the central part of the photograph, with clast size falling particularly rapidly beneath the darker mudstone layers. Stok Valley.

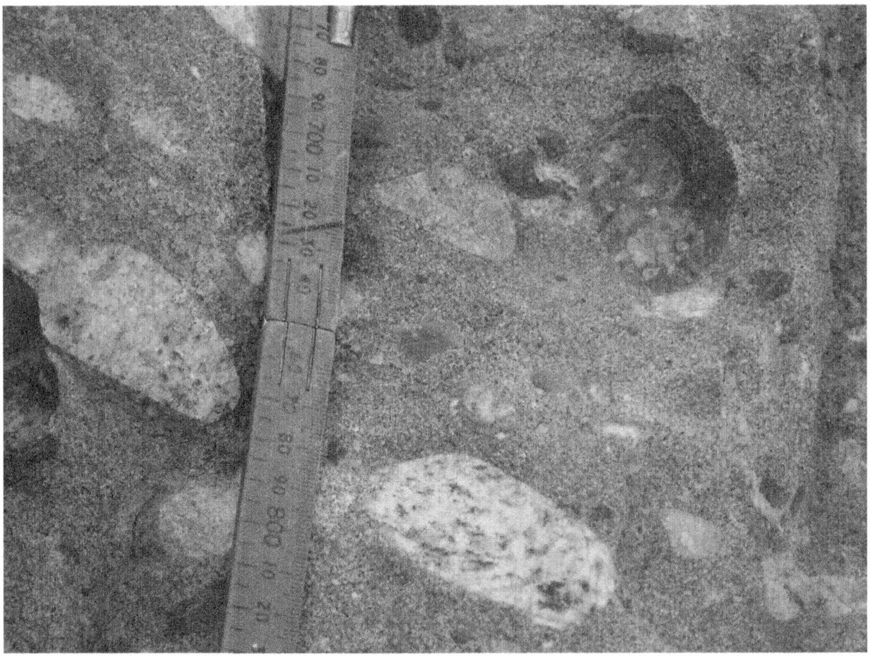

FIGURE 5.23 Indus Molasse heterolithic Conglomerate. Hemis. *Source*: Author's photograph.

around within the flow. Flows such as these are termed *debris-flows*. Abundant poorly sorted sediment is mobilised by high volumes of water, for example, rivers in flood. Rivers in this state can transport very large particles, including boulders, as their hydrodynamic power is increased by hundreds to thousands of times compared to normal steady-state conditions. The additional volume and velocity of rivers in flood are in a state of high momentum. Particles of very large size can only be transported when the rivers are in a high-energy state. Once the power of the rivers diminishes, the boulders and blocks rapidly deposit.

Figure 5.24 exhibits the transition between conglomerate and pebbly sandstone facies. The larger clasts are rounded to well-rounded, with a maximum size of around 4–7 cm. Long axes of the larger clasts are all oriented in the same direction: parallel to the bedding. A variety of lithologies is present within the rock, including granite, volcanic rock, quartzite, and diorite. The larger clasts are concentrated within one bed of Figure 5.24. This bed has a high concentration of pebbles, gravels, and coarse sand, indicating rapid deposition from a hyper-dense fluvial current. Large clasts are surrounded by a coarse-grained gravel-rich matrix. Lenses of coarse gravels surrounded by sandstone indicate rapid deposition around an obstruction or in the current lee-side of an obstruction. The larger pebbles exhibit coarse-tail inverse grading: they fine downwards, which may indicate that the rocks are 'the wrong way up' or deposited from a hyper-dense current. The upper part of Figure 5.24 contrasts with the lower part, with only a small number of clasts larger than 1 cm. This unit is poorly sorted with larger clasts of sub-angular quartz, granite, and volcanic material, but is more matrix-dominant. The bulk of the rock comprises grains of gravel and coarse sand size, dominated by quartz, with numerous lithic fragments. Sub-angular to sub-rounded clasts are the most common clast shapes. The existence of two quite different facies adjacent to each other indicates quite different flow regimes. A hyper-concentrated, high-energy flow transported and deposited the conglomeratic facies, whilst a high-energy but lower-density current deposited the gritty sandstone. These rocks were deposited more distally (further from the source region) than the conglomerates presented in Figure 2.91. The graded nature of the units suggests a turbiditic or braided river origin.

Figure 5.25 displays a different coarse-grained, poorly sorted rock to the examples above. The more angular clast shapes indicate that this rock type is mostly a breccia (a conglomerate with angular large clasts). Large clasts have high length/width ratios: their shape is lenticular. The long axes of the large clasts are aligned parallel to bedding. Fluvial currents align clasts of this shape, with high length/breadth ratios. The hydrodynamic nature of clast shapes affects their behaviour within water currents. Rounded or sub-rounded spherical particles behave quite differently from lenticular particles. The former are more difficult to retain in suspension: spherical clasts may be moved through the rolling of clasts along the river bed and/or pushing them into suspension for short periods of time (termed *saltation*). Lenticular clasts present large surface areas to water currents, allowing them to be retained in water suspension for longer periods of time. Clast alignment (termed *clast-imbrication*) suggests the presence of strong directional fluvial currents. Along with granite and volcanic clasts, this breccia contains abundant red silt/mudstone clasts, indicating a different sediment source region than other figures.

Figure 5.26 presents a panoramic view of an Indus Molasse Mountain ridge exposure. Tens of metres of dipping stratigraphy are exposed. The thickest unit crops out on the right-hand side of Figure 5.26: these are purple-thickly-bedded siltstones and mudstones, with interbedded thin sandstones. Stratigraphically

FIGURE 5.24 Alternating block and pebble-rich layers and coarse sand-rich layers, Indus Molasse. Hemis. *Source*: Author's photograph. 1cm of the metre ruler is exposed.

FIGURE 5.25 Breccia/Conglomerate Facies. Indus Molasse. Zanskar Gorge. *Source*: Author's photograph. Visible pencil is 15cm long.

FIGURE 5.26 Bedded purple siltstones and mudstones overlain by bedded brown-buff sandstones with thicker interbedded siltstones. Indus Molasse. Stok. *Source*: Author's photograph.

overlying this unit is a thick sequence of alternating resistant sandstones and conglomerates which form prominent ridges and benches. Purple and grey mudstones are interbedded with the sandstones. The change in character in the stratigraphy from right to left in Figure 5.26 indicates a change in facies and palaeo-sedimentary environments with time. Purple siltstones and mudstones are indicative of quieter sedimentary environments. They were deposited within a large and long-lasting lake. The thickness of the deposit suggests that it took a reasonable amount of time to accumulate this magnitude of sediment. A change to abundant thin- to thickly bedded sandstones indicates a change to higher-energy fluvial environments. Interbedded red-purple-grey mudstones are suggestive of quieter environments, such as floodplain mud deposits within a river floodplain, or the presence of ephemeral lakes and standing bodies of water. Regular intercalations of sandstone and mudstone suggest a cyclicity of environments with higher-energy fluvial channel environments dominating, punctuated by widespread floods across fluvial plains.

Figures 5.27 and 5.28 present outcrops of folded and thrust-faulted Indus Molasse east of Stok, south side of the Indus Valley, south of Leh. Figure 5.27 presents a geological interpretation of the main structures, placed next to the original image. A range of contorted and folded strata can be discerned. Fold styles vary from tight, upright, to inclined isoclinal shapes, to more open folds. Figure 5.28 indicates the movement of a thrust fault towards the north-northeast, placing the rock mass on the right-hand side over the rock mass on the left-hand side. Rocks on the left-hand side of the lower photograph are extensively deformed and folded, which contrasts with less folded/deformed rocks on the right-hand side. The Indus Molasse records a long period of deformation as the Ladakh, Karakoram, and Indian plates collided with one another over tens of millions of years. The Indus Molasse is predominantly thrust towards the north. Folded strata are common. Thrust faults and intra-formational faults break up the Indus Molasse strata and the continuity/coherence of folds. Fold structures are sheared along their limbs, reflecting extensive, protracted, and complex crustal movements over time.

Figure 5.29 presents a panoramic photograph, looking southwards, of Indus Molasse mountains. The plains are at an altitude of c. 3,400 m above sea level (asl) with the highest peaks at over 5,000 m asl. This winter scene allows bedding planes to be picked out by snow. Bedding dips towards the south but varies from south-west to south-east. A number of thrust units dissect the mountains and move crustal units above the thrust faults to the northeast, north and northwest. As tectonics move crustal fault blocks in northerly directions, they tilt the fault blocks above the thrust fault southwards. Plains in the foreground slope from the mountains to the River Indus, composed of variable deposits, mostly coarse pebbly sandstones, of fluvial and alluvial origin.

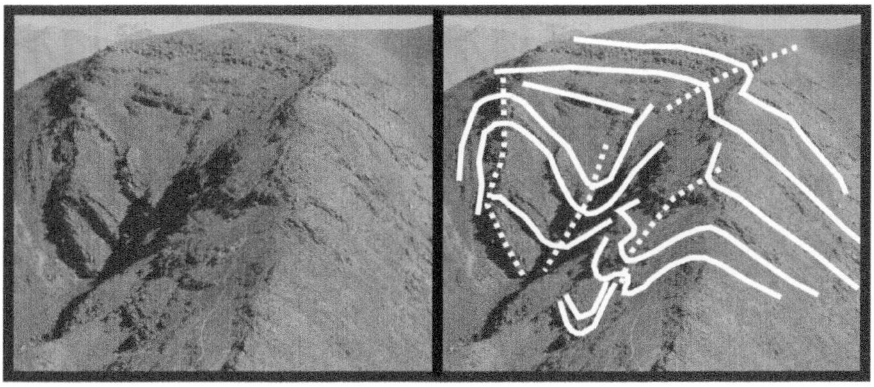

FIGURE 5.27 Folded and thrusted Indus Molasse sandstones. *Source*: Author's materials. Note: Tight isoclinal antiform with an axial plane dipping towards the right. East Stok.

FIGURE 5.28 Folded and thrusted Indus Molasse. *Source*: Author's materials. Note: North-Northeast to the left-hand side. Continuous lines indicate strata, pecked lines indicate fold axial planes, and the red continuous line indicates a thrust fault plane. East of Stok.

FIGURE 5.29 Indus Molasse mountains around Matho, with extensive alluvial plains. *Source*: Author's photograph.

FIGURE 5.30 Views across the Indus Plain, to the north (left-hand side) and to the south (right-hand side). *Source*: Author's photographs. Note: The left-hand side view looks across the capital town of Leh to mountains composed of the Ladakh batholith. The right-hand side photograph looks across the Indus plains to the Indus Molasse mountains.

Figure 5.30 places a northerly view against a southerly view across the wide and extensive Indus Plains close to Leh, the capital of Ladakh. Rounded mountains composed of the Ladakh batholith on the left-hand side contrast with the more rugged and serrated mountains of the Indus Molasse. Snow picks out the bedded nature of the Indus Molasse Rocks and the contrasting unbedded crystalline, plutonic rocks of the Ladakh batholith. We now leave the Indus Molasse behind and move northwards to the geology of the Ladakh batholith.

LITHOLOGIES, FIELD OUTCROPS, AND LANDSCAPES OF THE LADAKH BATHOLITH

The Kohistan-Ladakh batholith crops out over the bulk of the Kohistan-Ladakh island arc (Figure 5.31), extending some 6–700 km along strike throughout Afghanistan, Kohistan and Ladakh, with an outcrop width up to 100–130 km (Figure 5.31). Batholiths extend further east in the Lhasa terrane within southern Tibet: If all the batholiths are combined along the whole Trans-Himalaya-Himalaya region, they extend for some 2,500 km along strike, trending NW-SE within Kohistan-Ladakh and more east-west in Tibet. The Kohistan-Ladakh

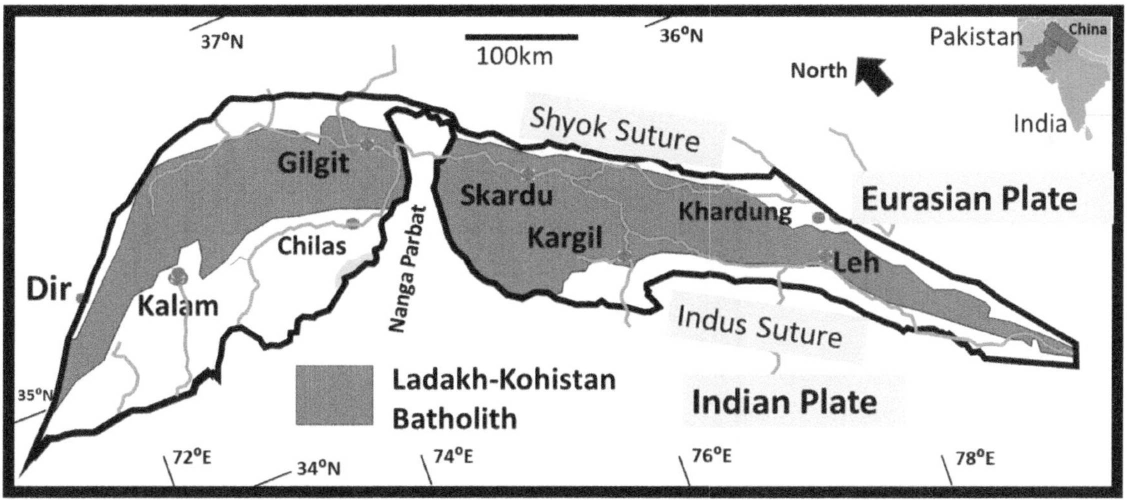

FIGURE 5.31 Geological map of the Kohistan-Ladakh batholith. *Source*: Author's map. Note: Inset shows the location of the map straddling India, Pakistan and Tibet/China.

batholith structure has been estimated to be between 30 and 8 km thick, with the thickest part of the batholith situated in Kohistan. The batholith is aged between Jurassic and Oligocene, or from around 150 Ma to 30 Ma. Around Leh, the batholith is dated between 70 and 45 Ma. The oldest units within the batholith, dated at 100–150 Ma, are situated between Skardu and Kargil, and north of Gilgit (Petterson, 2010, 2019, 2023).

Figure 5.32 is a diagrammatic cartoon of the structure of a batholith. Batholiths can be huge crustal structures. The Kohistan-Ladakh batholith is around 600 km long × 100 km wide × 10–30 km deep within the crust. Batholiths are comprised of a myriad of individual magmatic intrusions. Intrusions refer to the nature of the bodies: large to moderate volumes of magma rise in the crust, 'intrude' into, and freeze within, pre-existing crust. Magmas rise within the crust according to the difference between magma density and the surrounding crust. Extensional tectonics assist the rise and intrusion of magmas, providing space, fractures, and conduits for magmas to ascend. Magmas will struggle to rise higher into the crust when the surrounding crust is of the same/similar density as the magma. Whilst there is a significant density contrast between the magma and the surrounding crust, the magma has the relative buoyancy, momentum, and potential to rise, following Stokes' law of physics. Batholiths comprise rocks of widely varying composition from gabbro to diorite through to granite. The bulk of batholiths are composed of diorites, granodiorites, and granites. The shape and size of individual intrusions can vary enormously. Larger intrusions are called plutons and can extend over kilometres to tens of kilometres. At the other extreme, sheet-like intrusive bodies called dykes and sills may be only millimetres or centimetres to metres in width, although they can extend for kilometres along their length. The overall composition and density of continental crust is strongly influenced by batholiths, which can comprise the bulk of the crustal volume.

Batholiths form much of the middle continental crust. Continental crust has an average density of 2,700 kg/m^3, significantly less than that of oceanic crust (3,000 kg/m^3) and that of the Earth's mantle (4,500 kg/m^3). This density contrast has a strong influence on tectonics. It is difficult to subduct large quantities of continental crust back into the mantle, particularly when the crust thickens to over 17 km. Most continental crust is 30–35 km thick on average. Whilst small percentage volumes of the edges of continental plates may be subducted, and sediment derived from continental crust can be subducted back into the mantle, the great bulk stays at the Earth's surface. The age of continental crust is much older than that of oceanic crust. Continental crust can be over 4 billion years old (Wilde et al., 2001), whilst the oldest oceanic crust is mostly 180–200 Ma, although 340 Ma oceanic crust is preserved in the Mediterranean Sea (Granot, 2016). Oceanic crust is continually recycled back into the mantle, whilst continental crust, once created, becomes a permanent entity of the Earth's surface.

Continental crust has grown over geological time. At the time of the early Earth from 4.56 to c. 4 billion years, primitive higher-density, basaltic crust dominated the Earth's surface. With the maturing of

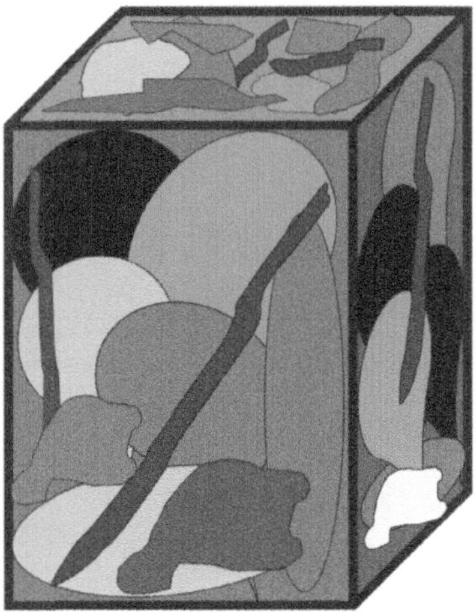

FIGURE 5.32 Schematic cartoon representative of a section through a batholith. *Source*: Author's diagram. Note: Batholiths are made up of magmatic intrusions of all compositions, shapes and sizes, from tear-drop and balloon-shaped intrusions to pipe-like dykes and amorphous intrusions. The section above could be up to 30–40 km thick, but mostly 7–15 km thick.

tectonics, increasing amounts of batholithic material were produced, and the continental crust grew over time. Continental crust now forms c. 30% of the Earth's surface (Hawkesworth et al., 2019.)

Batholiths are produced above subduction zones, either as oceanic crust subducting beneath another piece of oceanic crust, forming island arcs such as the Solomon Islands, Vanuatu, or the South Sandwich Islands, or, in the case of oceanic crust subducting beneath continental margins, such as in the case of the Andes mountains, where the East Pacific and other oceanic plates subduct beneath the South American continent (Marshak, 2019, chapter 6). Subduction causes melting to occur within a mantle wedge zone above the subducting plate. Volatiles emanate from the subducting plate, lowering local mantle melting temperatures. Frictional heat is added to the heat budget, raising local mantle temperatures. As the mantle partially melts, it produces basalt/gabbroic magmas that begin to rise, according to their relative density and buoyancy. When magmas rise, they crystallise different minerals such as olivine, feldspar, pyroxene, and amphibole. Magmas evolve in composition as they rise, becoming increasingly silica-rich, producing dioritic, granodioritic, and granitic plutons.

Figures 5.33, 5.34, and 5.35 illustrate different panoramic views, looking northwards from the Indus Valley across the Ladakh Mountain Range, composed of the Ladakh batholith. Batholith mountain morphologies can be more rounded, as compared with mountains composed of bedded Indus Molasse or the Upper Zanskar Sequence. Plutonic rocks are crystalline and don't have prominent bedding structures, as in sedimentary rocks. Glaciation can produce sharper topography, including some of the knife-edge upper ridge crests. Figures 5.33/34/35 provide views of large swathes of the Ladakh batholith, extending over kilometres horizontally and 1,000–1,500 m vertically. Mountain exposures vary in colour from dark to light. Granitic and granodioritic rocks are lighter coloured (*termed leucocratic*) whilst gabbros and gabbroic diorites are dark coloured (*termed melanocratic*) with diorites in-between (*termed mesocratic*). Granitic rocks can be seen to 'cut' (mostly) the darker diorites and gabbros. The relative age of plutonic rocks is determined by geologists in the field who study cross-cutting relationships. The oldest plutonic rocks are cut by everything, whilst the youngest plutonic rocks are cut by no other plutons and cut across all other (older) intrusions. These relationships provide data for determining the relative ages of batholith units. Figures 5.33/34/35 all present the relationship that granites are (in general) younger than the gabbros and diorites. Igneous rocks, particularly granites, can be dated by radiometric dating methods such as Uranium-Lead dating of the mineral zircon, which concentrates Uranium, and is a common accessory (minor) mineral in granites. The shapes of plutons in Figures 5.33/34/35 are highly variable. Some are irregular. Some are very large and occupy much of a mountainside, making it difficult to assess their shape, although they must be tens of cubic kilometres in volume. Other bodies are sheet-like, with a sub-horizontal disposition, termed *sills*. Granitic sills vary in thickness from a few metres to hundreds of metres thick and can extend right across a mountain. Occasional vertical sheets (dykes) are visible.

FIGURE 5.33 View looking north from Spituk Gompa towards the Ladakh Himalaya, north of Leh. *Source*: Author's photograph. Note: Highest snow-capped mountains are between 5,000 and 6,000 m asl. Much of the mountainscape comprises granitic (lighter coloured) and gabbro-gabbroic diorites (darker) rocks. Many plutons here are horizontal sheet or sill-like in nature. Mostly, the granitic plutons cut the gabbro-diorites.

FIGURE 5.34 View looking northwards towards the head of Taru Valley. *Source*: Author's photograph. Note: Large granitic plutons form the bulk of the rounded mountains. Sharp arête-ridge tops were excavated at the height of recent glaciations.

FIGURE 5.35 Valley north of Stakna Gompa, displaying a strong mixed-batholith character. *Source*: Author's photograph. Note: Darker gabbro-diorites are intruded by granitic sheets (sills) and plutons, whilst some granitic bodies are intruded by later basaltic dykes.

Figures 5.36 and 5.37 illustrate plutonic contacts, where one intrusion cuts another and forms a pluton-pluton contact. In Figure 5.36, a coarser-grained gabbro cuts a fine-grained gabbro. The contact is rather diffuse and ambiguous: not clear-cut to interpret. Close inspection reveals how the coarser-grained gabbro has smaller crystals close to the intrusive contact, suggesting that this is a *chilled margin* where the later hotter magma cools against the cooler host rock it intrudes, diminishing crystal size through more rapid cooling.

Figure 5.37 presents a clear-cut intrusive margin with a fine-grained basaltic dyke clearly cross-cutting a coarse-grained granite. The basalt has a well-developed chilled margin, with a 2–5 mm wide boundary zone occupied by rapidly chilled crystals. The main body of the basalt presents a *porphyritic texture* formed from larger (*phenocryst*) crystals set within a finer-grained crystalline matrix. Square-shaped, 5 mm-sized, black pyroxene phenocrysts are visible. The granite is a coarse, equigranular-textured plutonic rock, composed of white feldspar, glassy-vitreous quartz, and thin sheet-like micas: both dark (biotite) and white (muscovite) micas are present. Crystals are up to 15 mm in length.

Figures 5.38 and 5.39 present medium-coarse-grained granites/granitoids. The term *granitoid* is used for a granite-like rock where there is an element of uncertainty as to the precise classification of the rock. Figure 5.38 presents a coarse-grained granite, with the larger crystals measuring 5–10 mm. The rock has an equigranular texture with most individual crystals displaying similar sizes. The crystalline shape of constituent crystals is apparent from the regular geometric nature of many crystals. Squares and rectangles are common shapes. Interlocking networks of crystals are clearly visible. White crystals are feldspars, glassy crystals are quartz, and dark minerals are biotite and amphibole. This granite has taken on a tectonic planar cleavage fabric, particularly visible close to the pencil. Figure 4.39 presents a felsic granite with larger and more abundant feldspar and quartz. Dark biotite mica minerals are finer-grained and elongate in shape. The coarse-grained nature of these rocks indicates slow cooling of magmas. For this to occur, magmas reside within a crust with a warm-hot ambient surrounding temperature, or are such a large volume that it takes considerable time for magmas to cool. Surrounded by insulating (warm-hot) host rock crust, magmas can take hundreds or thousands of years to cool and freeze, allowing time for crystals to grow and develop well-formed (*euhedral*) crystal shapes.

Figure 5.40 presents a different granite facies from the previous examples above. The key difference is the grain size. Figure 5.40 granite is fine-grained, with an average grain size of 1–2 mm, and even finer-grained crystals within the microcrystalline matrix. The crystalline nature of fine-grained granites

FIGURE 5.36 Two contrasting gabbros in contact. Sankar. *Source*: Author's photograph. The right-hand side gabbro is finer grained, whilst the left-side gabbro contains crystals over 1 cm in size. Visible pencil is 15cm long.

FIGURE 5.37 Coarse granite intruded by dark basaltic dyke. *Source*: Author's photograph. Note: Leucocratic granite composed of feldspar (white), quartz (glassy-vitreous) and biotite-muscovite micas measuring up to 2 cm in length. Melanocratic basalt is finer-grained with a porphyritic texture. Square pyroxene phenocrysts measure up to 5 mm. Note how the basalt grain size reduces against the granite contact. Phyang.

is less apparent to the naked eye, although this can be observed with a hand lens or magnifying glass. This granite has a milky-white appearance, with an abundance of quartz and feldspar crystals dominating the appearance. Dark minerals (mainly biotite mica) make up less than 10% of the rock mass. Silvery muscovite mica grains catch the light. Granites highly abundant in felsic minerals are classified as *leucogranites*. Technically, leucogranites refer to granites with less than 5% dark minerals: the granites in Figure 5.40 do not fall strictly within the leucogranite classification but have similar characteristics. Leucogranites form through partial melting of continental crust, or from the extreme end product of magmatic crystallisation (as magmas crystallise, the residual liquid becomes increasingly felsic). As magmas rise through the continental crust, they absorb/assimilate surrounding crustal materials and, on occasion, melt crust adjacent to magma chambers. The fine-grained nature of Figure 5.40 leucogranite indicates a more rapid cooling of its parent magmas compared to the case of Figure 5.38/9.

Figures 5.41 and 5.42 exhibit rocks of a more granodiorite composition. There is a greater abundance of dark minerals compared to Figures 5.38–5.40. Feldspar (white) and amphibole (dark) minerals are the largest and best-formed crystals. Feldspars display square and rectangular shapes. Amphiboles display square-rectangular shapes along with lenticular and extended hexagonal-octagonal shapes. Quartz has a blue-pale vitreous colour, with a range of shapes, from pseudo-rectangular to more amorphous. Quartz is the last mineral to crystallise and occupies spaces left by the crystallisation of other minerals. All crystals need space to form crystal shapes. If crystals compete for space, crystal shapes

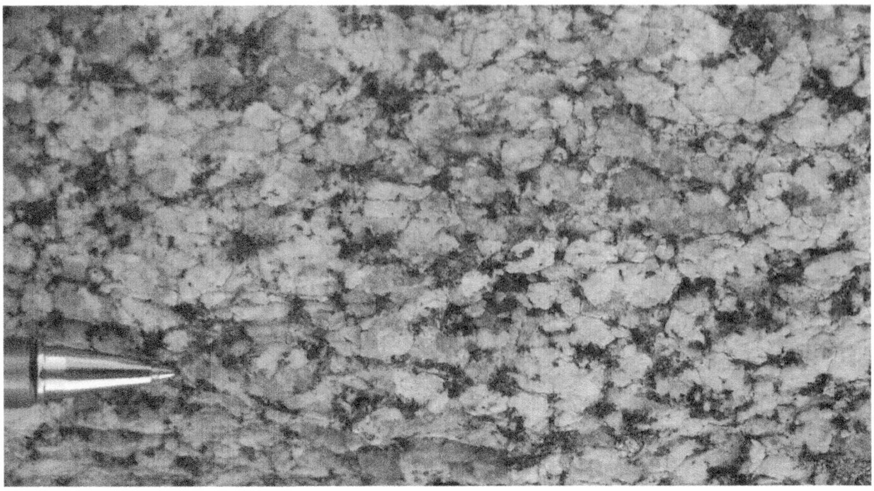

FIGURE 5.38 Coarse granular granite from Taru. *Source*: Author's photograph. Note: Leucocratic minerals are dominated by white feldspar with glassy quartz crystals. Darker biotite and amphibole crystals are present. This granite has experienced deformation and takes on a cleavage. Taru, west of Leh. 2.5cm of a pencil is visible for scale.

FIGURE 5.39 Coarse crystalline granite dominated by feldspar and quartz with subordinate biotite and amphibole. *Source*: Author's photograph. Pencil is 15 cm long.

struggle to form geometrical euhedral shapes, and minerals take on more amorphous appearances. Both photographs contain *xenoliths* or 'foreign bodies'. These are ovoid shapes at the centre of both photographs. Xenoliths are not part of the primary magmatic story. They form by falling into the magma chamber or by being forcefully 'acquired' by magmas ripping off pieces of crust in which they are intruding. Xenoliths are chemically and physically changed as they react with the host magma. Both xenoliths display a core facies at the heart of their entities, which has been less affected by the host magma. At the contact zone between the xenolith and the magma, xenoliths are finer-grained and grow

FIGURE 5.40 Fine-grained milky-white leucocratic granite, with a small abundance of melanocratic minerals. Taru, west of Leh. *Source*: Author's photograph. 3cm of a pencil is visible in this image.

FIGURE 5.41 Granodiorite with around 20–25% melanocratic minerals. *Source*: Author's photograph. Note: Contains a dark basaltic 'nodule' or *xenolith* with sharp edges and is unassimilated. Ganglas, north of Leh.

new and different minerals. Often, xenoliths become totally absorbed, are 'melted-out', and fully incorporated within the host magma. The xenolith in Figure 5.42 has been somewhat assimilated and *metasomatized*. Metasomatic processes involve physicochemical reactions and volatile alteration, which alter the primary nature of the metasomatized body.

Figures 5.43 and 5.44 display examples of a diorite and a gabbro, respectively. An increasing abundance of dark minerals characterises these more mafic rocks, compared to granites and granodiorites. The gabbro has roughly 50% dark minerals and 50% light minerals. Both figures exhibit equigranular coarse-grained textures. The diorite has larger crystals measuring 7–8 mm and the gabbro up to 12–15

FIGURE 5.42 Granodiorite with long lenticular (black) amphibole crystals. *Source*: Author's photograph. Note: The central xenolith has been metasomatically altered and somewhat assimilated. Ganglas, north of Leh. 8cm of a pencil is visible in this image.

mm. Quartz is much less abundant, or non-existent, within the gabbro. Gabbros and diorites are closer in composition to the 'primary magma' that results from mantle melting above subduction zones. . The gabbro in Figure 5.44 exhibits a recrystallised and deformed texture. Crystals are stretched and aligned, forming an incipient metamorphic foliation fabric. Dark minerals and light minerals separate, forming alternate melanocratic and leucocratic layers within the rock. The coarse-grained nature of the gabbro may be, in part, due to additional metamorphic processes that have acted upon the igneous rock after it cooled and fully crystallised. The Ladakh batholith has been caught up within the deformation events of the Himalayan orogenesis in places, creating strong metamorphic fabrics, shear zones, and thrusts locally (Petterson, 2023).

Figure 5.45 exhibits a diorite (micro-diorite) dyke intrusion, west of Leh. This dark dyke tracks across the mountainside, cutting everything else within the Ladakh batholith locally. The dyke must be the youngest intrusive event within this part of the batholith. Dykes tend to be sheet-like in shape, in three dimensions. They are (mostly) relatively thin, but can extend for many kilometres across the country. Dykes are vertical to high-angle intrusions. Horizontal sheets are termed sills. This dyke is 1–2 m wide and is clearly visible for hundreds of metres across the mountainside. Ladakh dykes mostly vary in composition between basalt/microgabbro and diorite/micro-diorite, and are late in the magmatic story. When the prefix 'micro' is added to a plutonic rock name, it indicates a *hypabyssal* (shallow) intrusion origin, where crystal sizes are normally less than those of deeper-seated plutonic rocks. Terms can be utilised loosely with dyke and sill rocks. Volcanic terms such as basalt and andesite are sometimes applied. To intrude the batholith, dykes follow fractures that form a pathway for the magma. The dyke magma may also intrude at high pressure, forcing fractures to open up at the front edge of the magma. Extensional stresses within the crust at right angles to the direction of the dyke assist dyke magma transport. Figure 5.45 exhibits dyke orientation varying along strike with sudden 'jumps' in direction. This can happen for a variety of reasons, including magmas coming to the end of a particular fracture and 'jumping' to another nearby fracture. Or later post-intrusion tectonic faulting can alter dyke orientations.

Figures 5.46 and 5.47 are good examples of quartz-porphyry dyke lithologies. These rocks are excellent examples of a magmatic *porphyritic texture*. This term refers to the presence of large minerals within a matrix of smaller minerals. The larger minerals are termed *phenocrysts*. The largest feldspar phenocrysts

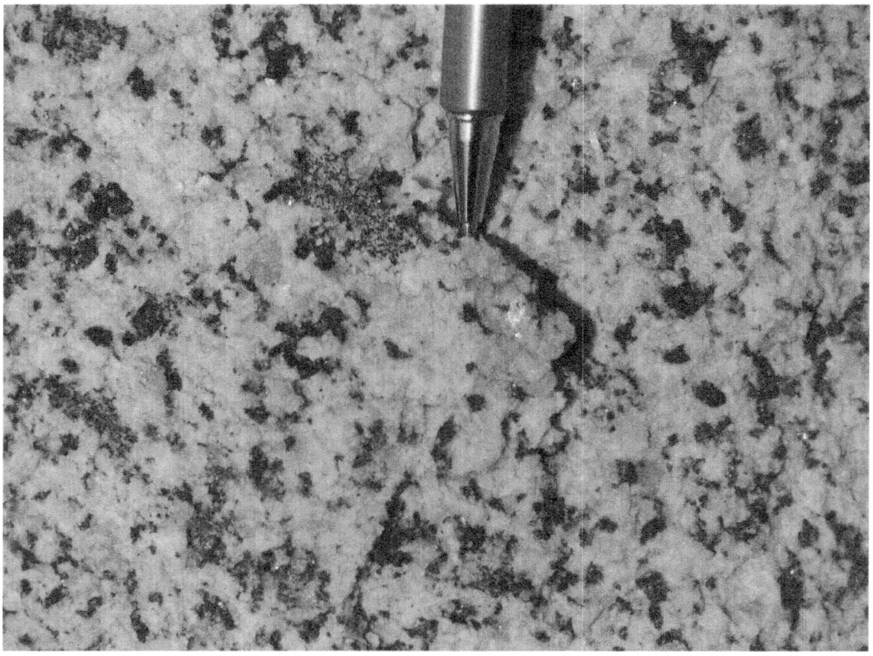

FIGURE 5.43 Medium-coarse grained diorite with c. 30% + melanocratic amphibole and biotite minerals. Leh. *Source*: Author's photograph. 3cm of a pencil is visible for scale.

FIGURE 5.44 Gabbro with 50% melanocratic and 50% leucocratic minerals. Leh. *Source*: Author's photograph. 12 cm of a pencil is visible for scale.

extend up to 1.5 cm in length. Glassy quartz phenocrysts extend up to 1 cm in length, with dark amphibole and biotite minerals being smaller in size. Phenocrysts are set within a fine-grained mesocratic (medium grey) microcrystalline matrix. A hand lens reveals the crystalline nature of the matrix. Many of the large feldspar phenocrysts, which are somewhat rounded in shape, are zoned. Bands of different colours are

FIGURE 5.45 Dark basaltic dyke cuts granitoid plutons west of Leh. *Source*: Author's photograph.

FIGURE 5.46 Quartz porphyry dyke lithology. *Source*: Author's photograph. Note: Large zoned, rounded, and rectangular feldspar phenocrysts occur alongside smaller quartz, mica, and amphibole phenocrysts, set within an andesitic/micro-dioritic, mesocratic, crystalline matrix. Taru, west of Leh. 2.5cm of a pencil visible for scale.

visible parallel to the edge of the crystal. This banding reveals a change in the chemistry of the crystals from the core to the rim. To grow large crystals such as these feldspar phenocrysts takes time. As conditions change and the chemistry of the surrounding magma evolves, crystal chemical compositions change, following the chemistry of the host magma. Banded colour changes within the crystals express this change to the naked eye. Phenocrysts include small dark crystals within their crystal form. As crystals grow, the phenocrysts 'swallow up'/incorporate/assimilate smaller crystals that are nucleating close to the evolving phenocryst.

Figure 5.48 presents an example of a felsic dyke intruding a host gabbroic-diorite. The milky-white dyke is a leucogranite with less than 5% melanocratic minerals. The dyke is fractured at right angles to its length, with its width exhibiting thickening and thinning, referred to as '*pinch and swell structure*'. The fractures and pinch and swell structures are evidence that the dyke has undergone deformation, post-intrusion. Fractures extend into the host gabbro, linked to brittle faulting. Pinch and swell deformation usually takes place within ductile conditions. Alternatively, the dyke may thicken and thin according to syn-intrusion space and magma availability. Figure 5.48 clearly presents a 'geological story'. The gabbro-diorite intruded within the batholith first. The host pluton was later intruded by a leucogranite dyke. Both were affected by later tectonic brittle deformation. Dyke margins illustrate the 'chilled margin effect'. Excess silica/quartz has crystallised along the margin. Leucogranite dykes may have formed through local crustal melting and the activity of hot hydrothermal solutions.

FIGURE 5.47 Quartz porphyry dyke rock. *Source*: Author's photograph. Note: Larger rounded phenocrysts of quartz and feldspar set within a leucocratic micro-fine grained crystalline matrix. Taru, west of Leh. 2cm of a pencil visible for scale.

Figure 5.49 presents an example of a mixed lithology outcrop from Tsemo, Leh. These types of outcrop are interpreted by some researchers as examples of mixed magmas, or evidence of the simultaneous co-existence of two magmas of different composition (Petterson, 2023 and references therein). In Figure 5.49, there are clearly two rocks present of two different compositions. Both lithologies are plutonic rocks. Dark rocks are microgabbroic/basaltic in composition. Light-coloured rocks are granitic in composition. For two such different magmas to co-exist and mingle, they must have originated from different magma chambers and intermingled at one location within the batholith and island arc and/or continental crust. Microgabbroic clasts are highly angular, perhaps previously being part of a partly solidified dyke, and subsequently ripped apart by a highly convecting, semi-solid, granitic magma. Some microgabbro clasts change colour at the edges, suggestive of metasomatic/partial assimilation by the granitic magma. Other microgabbro clasts show no such alteration. Batholiths are complex, with only some aspects of the batholith exposed at any one time within outcrops. Examples such as these illustrate how different magma chambers were simultaneously active, undertaking different magmatic evolutionary paths within the Ladakh crust.

Figure 5.50 presents an example of a deformed gabbro. Shear zones, strong shear movements, thrusting, and deformation within the Ladakh batholith create metamorphic gneissic fabrics within some plutons. Figure 5.50 is a close up of rock fabrics characterised by strong foliation and preferred mineral orientations, presenting the separation of dark (melanosome) and light (leucosome) layers, turning original magmatic crystalline textures into metamorphic gneissic foliations. See Petterson, 2023 for detailed discussions of this aspect of the Ladakh batholith around Leh.

We now leave the Ladakh batholith and move into Nubra, to see the geology, structure, and topography of north-Ladakh, the Shyok Suture, and the southern part of the Karakoram Terrane. Figure 5.51 looks north-eastwards towards the Shyok Suture Zone and the Karakoram mountains.

FIGURE 5.48 Cleaved felsic dyke intrudes gabbroic-diorite. Leh. *Source*: Author's photograph. 20cm of a metre ruler visible for scale.

FIGURE 5.49 Angular dark fine-grained basalt dyke clasts within a granitic matrix. *Source*: Author's photograph. Note: Mixed basalt and rhyolite magmas? A case of magma mixing? Tsemo, Leh. 25cm of a metre ruler visible for scale.

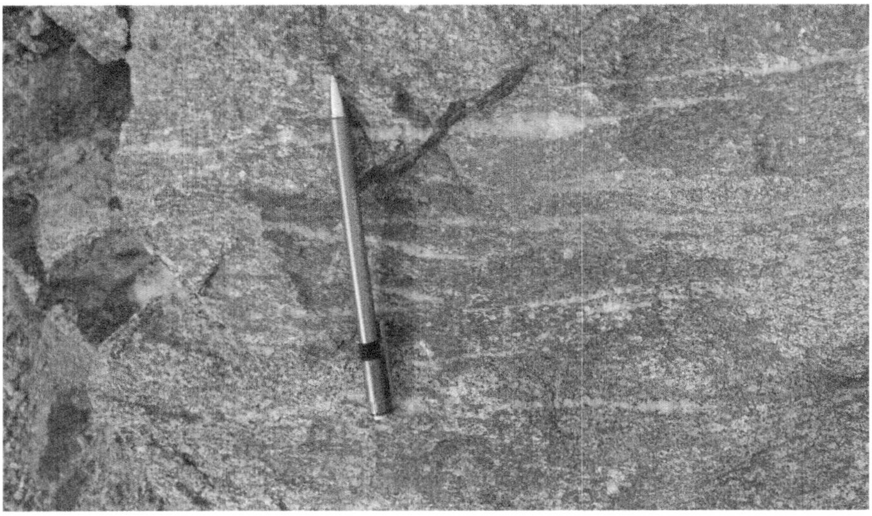

FIGURE 5.50 Foliated and sheared gneissic gabbro with quartz-feldspathic leucosome foliation layers and thicker amphibole-dominated melanosome foliation layers. *Source*: Author's photograph. Note: A significant proportion of the Ladakh batholith experienced deformation producing ductile-shear gneissic metamorphic fabrics. Leh. Pencil (15 cm long) visible for scale.

FIGURE 5.51 View from Khardung La looking north-eastwards towards Nubra, the Shyok Suture, and the Karakoram Mountains. *Source*: Author's photograph.

LITHOLOGIES, FIELD OUTCROPS, AND LANDSCAPES OF NUBRA, PANGONG, THE SHYOK SUTURE ZONE, AND SOUTHERN KARAKORAM PLATE

Figure 5.1 indicates the position of the Shyok Suture Zone (SSZ) and the Karakoram Fault (KKF). It is valuable to recap briefly the tectonics discussed in Chapter One with respect to this complex geo-tectonic zone. Parts of the area are outstanding examples of tectonic-dominated landscapes. Figure 5.52 beautifully illustrates how an understanding of tectonics and local geology opens up a deep interpretation of the topography. Here, the Karakoram Plate is visible on the right-hand side of the photo-mosaic, with the Ladakh Plate occupying much of the foreground and left side of the image. Suture structures are expressed by the sudden change in slope from mountain to plain. The active Karakoram Fault crops out on the right-hand side and separates the two plates. Other thrusts/faults comprise parts of the Shyok Suture

FIGURE 5.52 Interpreted view of the Nubra Valley area, North Ladakh, indicating positions of the Shyok Suture Zone (SSZ) and Karakoram Fault (KKF). *Source*: Author's photograph and interpretations. Note: Tectonics here have produced the largest and most extensive river valley plain in Ladakh. The Ladakh and Karakoram Plates meet here. The Main Karakoram Fault is the most active fault in the region and forms a suture between the two plates for much of its length. Other faults/thrusts also separate the plates within this complex Shyok Suture Zone (e.g. the pecked white line to the left of the Main Karakoram Fault).

Zone locally, including one that forms the base of another central mountain ridge in the image. The Shyok Suture Zone includes the whole image, incorporating rock units that are impacted within both plates as they collide. This zone has experienced a degree of extension, which has led to the local development of rifts within the mountain ranges. These rifts are infilled by sediment from rivers and ice and form the extensive Shyok Plains within the region of Nubra.

The Kohistan-Ladakh Island Arc (KLA) terrane is bounded to the south by the Indus Suture and to the north by the Shyok Suture (Figure 5.1). The geology we reviewed within the preceding sections examined the geology that has resulted from Ladakh being at one time an island arc, another time an Andean margin, and now, finally, a terrane within the Trans-Himalayan mountain chain. The Karakoram Plate/Terrane bounds the KLA to the northeast. In this book, we will briefly discuss the geo-tectonics of the Karakoram Plate: readers are welcome to explore more detailed accounts such as those presented in Searle (1991), Phillips et al. (2004), Treloar and Searle (2019), Pundir and Adhlakha (2024), and Saktura et al. (2026).

The Karakoram Plate/Terrane comprises NeoProterozoic-Palaeozoic basement sequences with multiple magmatic crustal additions from the Jurassic. Once part of the Gondwana Supercontinent, it rifted away from Gondwana during the Triassic and drifted across the Neo-Tethys Ocean, suturing with Asia during the Jurassic period. The Karakoram Batholith, aged between 130 Ma and 46 Ma (Jurassic-Tertiary), resulted from Andean-style subduction zone magmatism as the Kohistan-Ladakh Arc/Indian Plates subducted northwards beneath the Karakoram Plate/Terrane. Post-collisional deformation and magmatism caused widespread metamorphism and crustal melting, producing the Baltoro Plutonic Unit between 26 Ma and 13 Ma, and very young magmatism at <5 Ma.

The Shyok Suture, named after the Shyok River, visible in Figures 5.1 and 5.52, remains an active fault zone. The Main Karakoram Fault (MKF) is a continental-scale strike-slip fault. The term *strike-slip* refers to a fault that has a predominant horizontal motion: in this case, the mountains to the right of the fault have been moved towards the right (southeast) relative to the rest of Figure 5.52. The MKF is several hundred kilometres long, aligned NW-SE, extending into the Karakoram Terrane towards the NW, but forms the main plate boundary between Ladakh and the Karakoram locally.

Like the Indus Suture, the Shyok Suture has witnessed the disappearance of an ocean and major shearing and faulting as plates collide. Rock sequences found in this geo-tectonic setting reflect this environment, as was discussed previously with respect to the Indus Suture. Figure 5.53 captures a tectonic phenomenon that can be seen, touched, and tasted. The right-hand side photograph images a thermal spring surrounded by a grassy area within an otherwise arid mountain slope. White salt precipitates are visible. Thermal waters are piped and captured within a number of terraced baths, used for leisure activities and practical utilities, such as laundry. Active tectonic zones create geothermal waters locally. Frictional heating produced by colliding plates and moving faults, together with any magmatic heat that may be locally

FIGURE 5.53 Nubra thermal waters (right-hand side photograph) are piped and captured within terraced baths for local people and visitors to enjoy (left-hand side photograph). *Source*: Author's photographs.

FIGURE 5.54 The lower photograph presents a stripy mountain appearance composed of light coloured leucogranites and darker coloured metamorphic rocks including. Upper photograph presents the vast scale of the Pangong Leucogranite and Migmatite Complex. *Source*: Authors photograph.

present, elevates the ambient temperatures of groundwater. If the groundwater levels meet surface elevations, natural springs form. Large quantities of thermal waters can be used to generate electricity via geothermal power stations, and/or used for space heating in some countries. The presence of thermal waters here in Nubra, and elsewhere in Ladakh, is a testament to the dynamic nature of local geotectonics.

Figure 5.54 indicates the scale and vertical extent of the Pangong Migmatites and Leucogranites, extending for tens of kilometres horizontally and 1,600 m vertically. It has been estimated that these rocks have been transported upwards, up to 20 km, by the MKF, which also moved crust at least 150 km laterally (Searle et al., 1998). The active nature of the MKF is linked to the position of this part of Ladakh at the western edge of the Tibetan Plateau, a region of more intense local tectonics, caused by the underthrusting of India beneath Tibet. Rocks exposed here are evidence with respect to the nature of the continental crust within orogenic belts at a depth of around 20 km. Crust here was at temperatures high enough to at least partially melt crust (requiring a temperature of around 650–800°C at these depths and pressures) and produce high-grade gneisses. Pangong Migmatites have been dated at around 108–78 Ma with younger leucogranites dated at c. 41–23 Ma (Patel et al., 2009). Leucogranites form from the aggregation of highly silicic magma in sufficient quantities to allow the magma to move upwards through the crust. Leucogranites exposed in these localities, close to Pangong Lake, have a

FIGURE 5.55 Schists exposed on the south side of Pangong Lake. *Source*: Author's photographs. Note: 25cm of a 1m ruler exposed for scale.

sheet form with a large length/width ratio. They are several tens of metres thick and extend laterally for kilometres.

Figure 5.54 presents examples of exposures cropping out along the southern side of Pangong Lake. The upper photograph gives an idea of the scale of exposures which are common in this area, cut by road-builders. The lower photograph provides a close-up of the rocks. These rocks are at a much lower metamorphic grade than the Pangong Migmatites, comprising cleaved quartz-mica schists. The rocks are strongly sheared with well-developed schistose fabrics, typical of rocks caught up within suture zones and colliding plates.

A number of Cretaceous (115–72 Ma) volcanic and sedimentary units crop out within the Shyok Suture Zone (Figure 5.55) presents examples of volcanic rocks that crop out in Nubra. Within the Shyok Suture Zone, volcanic rocks have many similarities with the Dras Volcanics of the Indus Suture Zone

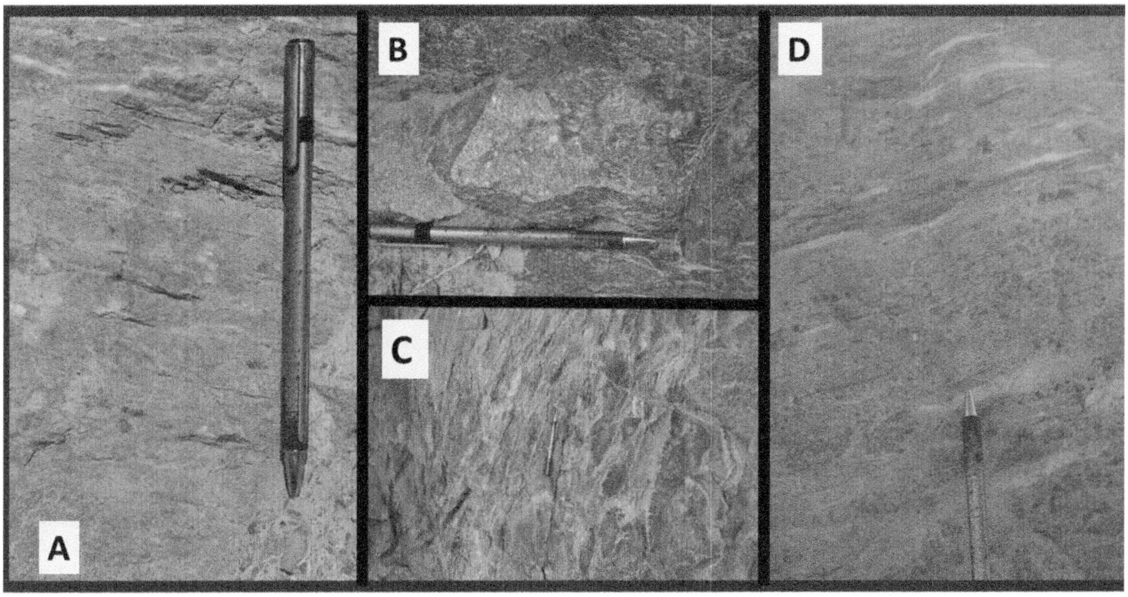

FIGURE 5.56 Shyok Suture Zone Volcanic Rock lithologies. *Source*: Author's photographs. See text for details. Note: Pencil in all photographs is 15cm long.

and the Chalt Volcanics of Kohistan. Volcanic compositions vary from basalt to rhyolite, with basalts and andesites dominant. Lavas and volcaniclastic rocks are present. Rocks can be highly sheared and deformed, with a strong cleavage fabric, which is to be expected within a suture zone. Metamorphism locally is mostly at greenschist grade: epidote and chlorite give many outcrops a green sheen. Shear zones, faults, and slickenside structures are common, indicating that numerous areas were caught up within the suture period deformation. Figures 5.55, B and C, present greenschist lava rocks. Figure 5.55, C presents a sheared outcrop with a number of thin volcanic units present. Figure 5.55, B is a porphyritic andesite, medium grey in colour, containing square and rectangular-shaped feldspar phenocrysts. Figure 5.55, A and D are pyroclastic volcanic rocks. A careful inspection reveals the presence of a streaky texture within the rock: a lenticular structure with high length/width ratios, some coloured black and others pale grey-white. The streaky clasts/particles are *fiamme* after 'flame' in Italian. These are squashed pumice relics. Aligned fiamme form a foliation called a *eutaxitic fabric.* Along with fiamme, the rock is composed of less- or undeformed crystals and clasts (composed of solidified volcanic material). The matrix is composed of ash, crystals, and glass shards, now recrystallised through metamorphism. These rocks are called *ignimbrites* and are *welded tuffs*. Welded tuffs are rocks composed of pyroclastic (hot volcaniclastic materials) clasts that were originally erupted at high temperatures and welded together as they accumulated within a deposit. Welding occurs at temperatures over 500ºC. Ignimbrites form from *pyroclastic flows* that erupt as hot (sometimes over 800ºC) mixtures of solid particles (crystals, ash, glass, rock fragments) and flow at low viscosities reaching eruption speeds of 700 km/h. Once they come to rest and accumulate, large volumes of pyroclastic particles weld. Large-scale pyroclastic flows form within island arcs and continents, subject to subduction. Ignimbrites and related pyroclastic flows are the most dangerous of all volcanic eruptions, as they can travel so fast and move across landscapes with relative ease. Shyok Suture volcanic rocks also contain facies indicative of oceanic eruption, including deep oceanic eruptions, alongside subaerial eruptions.

This chapter ends after a long traverse from the Indian Plate in Zanskar through an island arc in the Leh-Ladakh area, ending across the Shyok Suture within the Karakoram Terrane. Figure 5.56 presents a photograph across Pangong Lake to the terranes that now make up central Asia and Tibet.

FIGURE 5.57 Looking northwards across Pangong Lake to the Karakoram Mountains, the Karakoram Plate, and Eurasia. *Source*: Author's photograph.

REFERENCES

Hamilton, W. (1772). *Observations on Mount Vesuvius, Mount Etna, and Other Volcanos: In a Series of Letters, Addressed to the Royal Society*. London: Printed for T. Cadell, in The Strand.

Hawkesworth, C., Cawood, P. A., & Dhuime, B. 2019. Rates of generation and growth of continental crust. *Geoscience Frontiers,* 10(1), 165–175. https//doi.org/j.gsf.2018. 020004

Marshak, S. (2019). *Earth: Portrait of a Planet* (6th ed.). New York: W.W. Norton & Company.

Moore, J. G. (1975). Mechanism of formation of pillow lava. *American Scientist*, 63(3), 269–277.

Patel, R. C., Singh, S., & Padia, J. T. (2009). Evolution and chronology of the Pangong Metamorphic Complex adjacent to the Karakoram Fault, Ladakh: Constraints from thermobarometry, metamorphic modelling and U-Pb geochronology. *Journal of the Geological Society of India*, 166(5), 919–932.

Petterson, M. G. (2010). A review of the geology and tectonics of the Kohistan island arc, north Pakistan. *Journal of the Geological Society of London, Special Publication*, 338, 287–327. https://doi.org/10.5194/essd-14-4171.

Petterson, M. G. (2019). Plutonic and volcanic rocks of Kohistan, North Pakistan: Implications for island arc genesis and evolution. In P. J. Treloar & M. P. Searle (Eds.), *The Himalayas: A Modern Synthesis*. Special Publication of the Geological Society of London, No 483, pp. 123–164. https://doi.org/10.1144/SP483.4.

Petterson, M. G. (2023). *Himalayan Thick-Skin Basement Deformation of the Ladakh Batholith, Leh-Ladakh Region, NW India*. Cham, Switzerland: Springer Nature (Geology) Publications.

Phillips, R. J., Parrish, R. R., & Searle, M. P. (2004). Age constraints on ductile deformation and long-term slip rates along the Karakoram fault zone, Ladakh. *Earth and Planetary Science Letters*, 236, 305–319.

Pundir, S., & Adlakha, V. (2024). Geological evolution of the Karakoram Terrane since Neoproterozoic: A review of recent data. *Earth-Science Reviews*, 252, 104712.

Robertson, A. H. F., & Sharp, I. R. (1998). Tectonic setting of the Indus Suture Zone, Ladakh Himalaya. *Journal of Asian Earth Sciences*, 16(1), 1–15.

Saktura, R. B., et al. (2026). Magmatic evolution of the Karakoram Batholith (~160–62 Ma), Trans-Himalaya, NW India: Geochemical insights for arc dynamics and crustal growth. *Journal of Asian Earth Sciences*, 281, 106042.

Searle, M. P. (1991). Geology and tectonics of Karakoram mountains. *Geological Magazine*, 129, 358.

Searle, M. P., Cooper, D. J. W., Rex, A. J., & Fryer, B. J. (1988). Stratigraphy, structure and evolution of the Tibetan-Tethys zone in Zanskar and the Indus suture zone in the Ladakh Himalaya. *Transactions of the Royal Society of Edinburgh: Earth Sciences*, 79(2–3), 67–88.

Searle, M. P., Weinberg, R. F., & Dunlap, W. J. (1998). Transpressional tectonics along the Karakoram fault zone, northern Ladakh: Constraints on Tibetan extrusion. *Geological Society, London, Special Publications*, 135(1), 307–326.

Treloar, P. J., & Searle, M. P. (Eds.). (2019). *Himalayan Tectonics: A Modern Synthesis*. Geological Society, London, Special Publications, 483, 215–254. https://doi.org/10.1144/SP483.7.

Walsh, J. S., Reiners, P. W., Gehrels, G. E., & Pullen, A. (2019). Age and provenance of the Nindam Formation, Ladakh, NW Himalaya: Evolution of the intraoceanic dras arc before collision with India. *Tectonics*, 38(8), 2826–2849. https://doi.org/10.1029/2019TC005494.

Ladakh Tectonic- and Glacial-Dominated Landscapes

6

Although previous chapters have presented a range of landscapes with geological explanations, this chapter and the next focus on *landscapes themselves* and how they have been shaped through a range of *geomorphological* processes (natural landscape-shaping processes and phenomena). Sections in this chapter will highlight how tectonic and glacial processes form landscapes dominated by, or recording strong elements from, their respective activities. Beginning with tectonics, young mountain landscapes are especially instructive for visualising clear relationships between topographic shape/form and specific tectonic processes (Marshak, 2019, chapter 11). Faults are a more obvious tectonic agent that can produce linear mountain fronts, while folds shape mountains locally with their curvilinear morphology, and extensional tectonics allow valleys to form and accumulate large thicknesses of sediment.

Globally, tectonics have had, by far, the greatest impact on landscapes. Examples such as the existence of ocean basins and continents, continental-scale rift valleys, and the dramatic wall of Himalayan rock rising above the Indo-Gangetic Plain are all global-scale features. Figure 6.1 presents a view of a landscape with numerous causal influences, including the activity of tectonics and ice. The prominent central hill hosting Tsemo Gompa (temple) and Leh Palace is strongly indented by long, linear fractures that shape the mountain. The fractures, in turn, were produced by thrust-tectonic forces, creating well-defined escarpment features. High mountains in the distance are also shaped by strong fracture networks. Careful observation of the Tsemo Gompa Ridge shows that the ridge itself is an escarpment feature. Escarpments have three elements: a steep slope (termed a *scarp slope*), a lower-angle slope (termed a *dip slope*), and the *crest* of the escarpment, where the scarp and dip slopes meet. Figure 6.1 presents scarp-slope and crest features of the Tsemo ridge. The dip slope is largely hidden but can be deduced by the slope of the ridge beyond the crest, on which the Tsemo Gompa and the grey Leh Palace are built. Prominent fractures are clearly visible on the scarp slope, splitting the rock mass into extensive rectangular blocks. Fractures dip to the right, reflecting the dip of the escarpment. Fracture geometry is a clear dominant controlling element in ridge shape development. Small valleys are present on the face of the scarp slope with broken-up rock clasts, sand, and mud concentrated in these narrow, steep valleys. These zones are the products of water movement downslope during times of heavy rain and *mass wasting* of the slope, as slopes respond to gravity over time. The central ridge is surrounded by wide valleys on both sides, now occupied by the city of Leh. These valleys have been sculpted by glaciers, which carved broad 'U' shaped valleys. The movement of the glaciers is influenced by geological fractures and faults.

Figures 6.2 and 6.3 present two of the most extensive flat intra-mountain plain features in Ladakh. Figure 6.2 is taken from the village of Karsha, looking south-southwest towards Padum, Zanskar. Figure 6.3 looks north-westwards towards Hunder, Nubra, North Ladakh. The Padum Plain hosts three rivers: Tsarap, Stod, and Zanskar. The Hunder Plain hosts the River Shyok. River systems here contain abundant sediment loads and take advantage of space, producing multiple interconnecting channels with numerous river islands (typical of *braided rivers*). Both plains formed within the immediate proximity of major faults. In the case of Padum, the Zanskar Shear Zone bounds the southern side of the plain and produces a highly oriented linear mountain front, with the High Himalaya Range to the south. For the Nubra Plain, the Main Karakoram Fault and other Shyok Suture structures bound the plain to the north, and produce a linear southern mountain front

DOI: 10.1201/9781003794868-6

FIGURE 6.1 Tsemo Gompa and Leh Palace, Leh City. *Source*: Author's photograph. Note: Landscapes are the product of multiple geomorphological, geological, and tectonic processes. This figure presents a highly fractured escarpment, upon which the Tsemo Gompa (temple) and Leh Palace are situated. This escarpment is largely constructed through thrust-tectonics.

FIGURE 6.2 Wide, intra-montane plain resulting from tectonics and fault activity. *Source*: Author's photograph. Note: This photograph above is the largest plain in Zanskar which hosts Padum.

to the Karakoram Range. In the case of the Zanskar Shear Zone, the shear-fault structure has moved laterally as well as vertically, locally producing extension within the crust and facilitating a *rift valley* to form (Herren, 1987). The Main Karakoram Fault has a strike-slip (horizontal) motion component in addition to vertical motion (Searle et al., 1998). The same structures do not produce large valleys and plains continuously along their length. Larger plains are restricted to areas where tectonic structures have a significant horizontal motion component alongside a vertical component to their fault dynamics. Valley and plain morphologies evolve once produced by tectonics, shaped by ice, wind, and water.

Figure 6.4 is a view of Pangong Lake, one of the highest brackish water lakes in the world (4,225 m above sea level). This view looks south-eastwards. Pangong Lake is situated within the Karakoram Plate, to the north of the Kohistan-Ladakh Plate. Pangong is a drowned fault valley. Previously, this valley had been a river valley, but more recent fault movements led to the damming of the river and consequent lake creation (Searle et al., 1998). Figure 6.4 displays the strikingly linear form of Pangong Lake, here oriented

FIGURE 6.3 Extensive intra-montane plain, Nubra, resulting from active fault tectonics. *Source*: Author's photograph. Note: A wide, meandering, braided river is present in the foreground.

FIGURE 6.4 Pangong Lake, one of the highest brackish lakes in the world at over 4,000 m. *Source*: Author's photograph. Note: The Karakoram Mountains are visible on the far side of the lake. The south (right-hand) side of the lake parallels the Main Karakoram Fault.

NW-SE. In the far (right-side) distance of the photograph, a change in orientation is hinted at: as the lake suddenly alters direction to north-east-south-west: This lake shape is strongly influenced by geological structure. The Main Karakoram Fault is on the southern side of Pangong Lake, also orientated NW-SE, and extends north-westwards to the Shyok Valley, visible in Figure 6.3. The Karakoram Mountains are visible across the lake in Figure 6.4. A lake delta is present in the mid-foreground of Figure 6.4, produced by sediment eroded from the Pangong Range to the south-west (out of view, towards the lower right of Figure 6.4, discussed below in this chapter), and transported north-eastwards. All lakes are temporary features within mountain geomorphology. Eventually, they become infilled by sediment or drained if the natural dam that holds up the water is tectonically compromised.

The tectonic section of this chapter ends with an examination of Figure 6.5. This photograph looks south-south-eastwards down the River Tsarap Valley, southeast of Padum, Zanskar. The strong linearity

FIGURE 6.5 View south-eastwards down the River Tsarap Valley, southeast of Padum, Zanskar. *Source*: Author's photograph. Note: The strong linear nature of the High Himalaya mountain front is well demonstrated, along with the narrow river gorge formed in the distance. These geomorphological features are controlled by the Zanskar Shear Zone.

of the valley is clearly presented. The right-hand side of the valley is occupied by an evenly sloping mountain range, dipping towards the left (north-eastwards) at around 30°. This geomorphological feature manifests the dip slope of a high escarpment. The escarpment shape is, in turn, influenced by the dip of the dominant foliation within the High Himalaya Crystallines geology, which constitutes the mountain substrate. The Zanskar Shear Zone follows the trend of the River Tsarap Valley, or, more accurately, the river follows the trend of the Zanskar Shear Zone. The Zanskar Shear Zone (ZSZ) separates the Zanskar Sequence from the High Himalaya Crystallines (Chapter 3). The ZSZ is a major tectonic structure that extends throughout Ladakh, as well as linked to structures that extend throughout much of the Himalaya, including around Mount Everest, Nepal. The River Tsarap has taken advantage of weaker, highly sheared and fractured rocks within the fault zone. Water will always find its way through the weakest crust. The river is '*rejuvenated*' (has its erosive and carrying capacity powers increased) by the uplift of the mountains, caused by tectonism. Here, the River Tsarap has cut a deep, 'V' shaped valley, including numerous narrow valley gorge sections. Figures in this section demonstrate that the character, shape, and three-dimensional architectures of valleys and plains vary, linked to the precise nature of fault movements locally, alongside the combination of erosive agents such as glaciers, rivers, and wind. We now leave the theme of tectonics and move to the examination of the role of ice in shaping Ladakh.

LADAKH GLACIAL LANDSCAPES

Figures 6.6 and 6.7 present examples of modern-day Ladakhi glaciers, from Drang Drung, Pensi-La, Rangdum, and Nubra areas of Ladakh. These images illustrate some of the key features of glaciers and glaciers of differing size and extent. The glacier in Figure 6.6 has a source region at close to 6,000 m asl, with a terminal edge at around 5,000 m asl. Figure 6.7 glaciers have their source regions at elevations of 5500–6000 m and terminate as ice-dominated glaciers at around 5000–4400 m, depending on the size of the glacier and volume of ice. Figure 6.6 presents a clear glacier structure. Glaciers are sourced within snow and ice accumulation zones. As ice accumulates, it excavates a bowl or amphitheatre-shaped structure termed a *corrie*. Contours and ridges wrap around the accumulation zone of the glacier in Figure 6.6. As ice accumulates, it outgrows the corrie and flows downslope under the force of gravity. Glaciers rapidly erode steep valleys with flat valley floors, termed *'U' shaped valleys*. Whilst ice flows downslope, surface crevasses appear, formed through differential shear forces within the ice sheet. Glaciers move more slowly where they are in contact with rock and substrate. Central and upper zones move more rapidly.

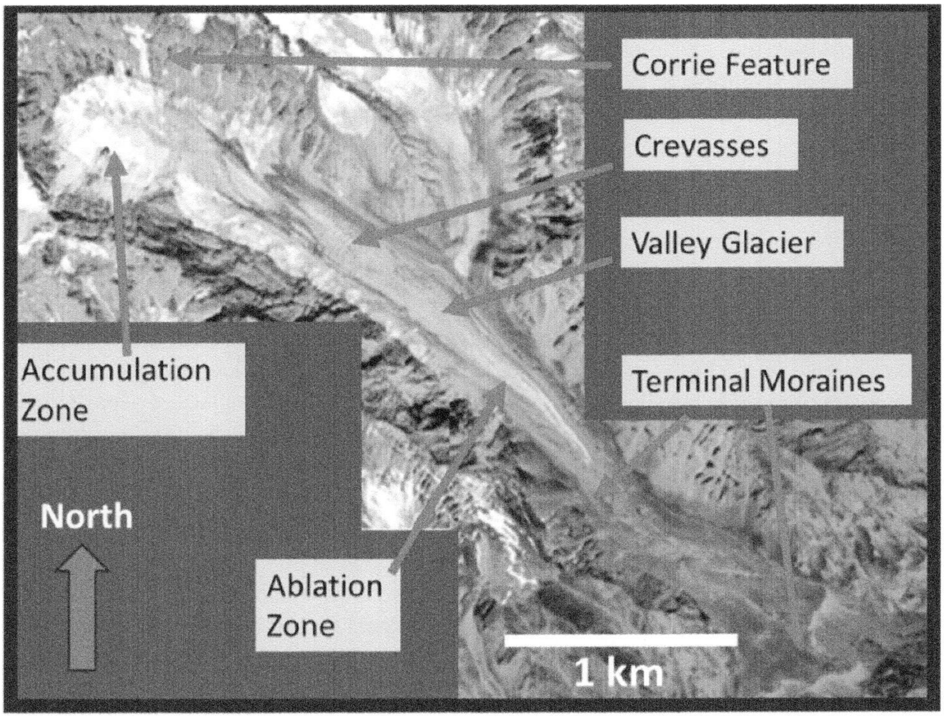

FIGURE 6.6 Glacier located west of Panamik Village and north of Hunder, Nubra, North Ladakh. Note: Glaciers extend to > 5,800 m above sea level. *Source*: Authors' work. Base Map acknowledgements. Google. (n.d.). Google-North-Ladakh). Retrieved February 13th 2025 from https://www.google.com/maps/search/panamilk+ladakh/@34.7919774,77.6001589,40317m.

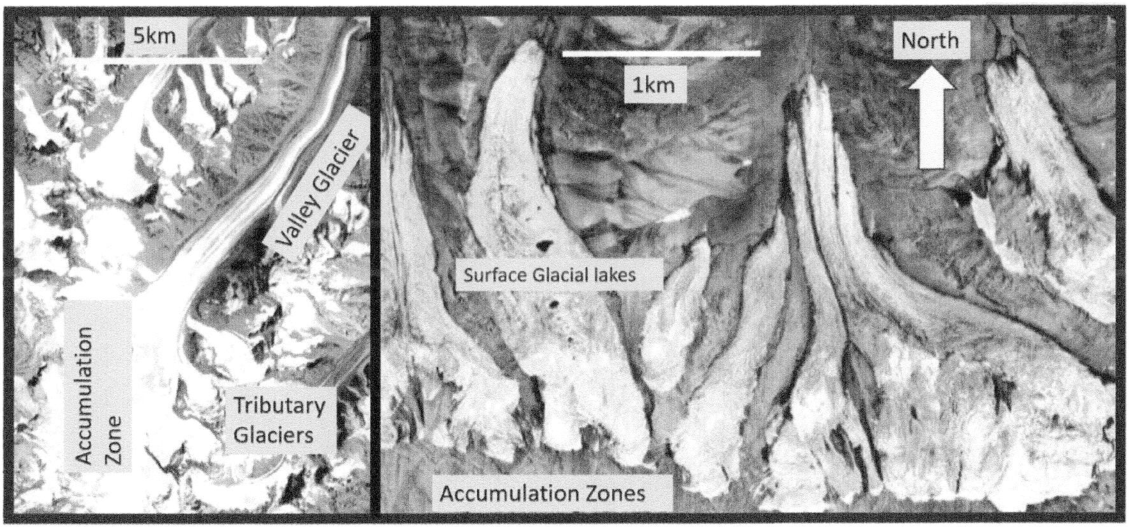

FIGURE 6.7 Glaciers from Drang Drung near Pensi La (left-hand side) and SE of Rangdum (right-hand side), Zanskar. Note: Highest mountains around the glaciers are almost 6,000 m asl around Drang Drung, and 5,600 m asl above SE Rangdum glaciers. Acknowledgements. Google.(n.d.). Google-North-Ladakh). *Source*: Base Map. Retrieved February 14th 2025 from https://www.google.com/maps/search/ Durung+Drung+Glacier/@33.7545505,76.1492685,40818m/ and https://www.google.com/maps/place/Pensi+La/@33.8680985,76.3450884,2548m

Velocity differentials within the mass of the glacier cause crevasses to form (Figure 6.6). As ice moves further downslope, it reaches a point at which melting exceeds accumulation, and ice begins to waste within the *ablation zone,* resulting in sediment deposition. Ice is a highly effective erosive agent, rapidly denuding the rock mass over which it flows. The terminal zone of a glacier is a zone of sediment accumulation, producing *moraines*, a generic term for glacial deposits. Over 1 km of terminal moraines are present in Figure 6.6, with the length of the glacier being around 2.5 km in total. For general background information on glaciers and glaciation, see Marshak, 2019, chapter 22.

Figure 6.7 displays glaciers of different scales. The glacier on the left-hand side (Drang Drung) is one of the larger Ladakhi glaciers, with a length of >20 km, and is regularly visited by travellers in Ladakh, as it is close to the road between Zanskar and Kargil. Glaciers on the right-hand side are smaller scale, at around 2–3 km in length. Drang Drung glacier has a relatively large accumulation zone of 25–30km^2, with smaller, tributary glaciers adding ice to the overall ice budget, and a tapering valley glacier form over its length. Darker linear zones are visible parallel to the length and side of the glacier: glaciers deposit moraines at the sides of the valley, producing *lateral moraines*. Smaller glaciers on the right-hand side of Figure 6.7 exhibit the main elements of any glacier, with an upper accumulation zone and a lower ablation zone. Fractures and lines of sediment are present parallel to the length of each glacier. Many glaciers have varying amounts of surface water, producing glacial lakes and pools. These smaller-volume glaciers may be under stress with rising temperatures locally. The ratio of ice accumulation to ice ablation is low, and ultimately these glaciers may disappear if local temperatures in the high mountain environments rise further.

Figures 6.8, 6.9, and 6.10 present images of glaciers at Pensi La and Pangong. Many classic features of glaciers are present in Figure 6.8. The accumulation zone is visible amongst the higher mountains in the

FIGURE 6.8 Drang Drung glacier, Pensi La, Zanskar. *Source*: Author's photograph. Note: Note the accumulation zone in the distance, structures within the main glacier body, the tapering form of the glacier, lateral moraines along the sides of the glacier, the ice becoming 'dirtier' towards its terminus/snout, and the glacial lake formed at the snout of the glacier.

FIGURE 6.9 High-mountain glacier formed within the Pangong Mountains, south of Pangong Lake. *Source*: Author's photograph.

FIGURE 6.10 Panoramic view of the Pangong Mountains, raised by the Main Karakoram Fault to heights in excess of 6,000 m asl, forming a host to a range of small to medium scale glaciers. *Source*: Image created from author's photographs.

distance: the greater thickness of the glacier is noticeable in this upper region. As the glacier flows, it produces linear structures parallel to its length and sides. Large volumes of dark-coloured lateral moraines are present between the edge of the glacier and the sides of the adjacent mountains. Lateral moraines have higher central ridges forming glacial levees. The colour of the ice changes towards the terminal point or *snout* of the glacier. The snout region is dark-coloured. Ice is present but is buried beneath sediment. The ratio of sediment: ice increases until a point is reached where the glacier has visibly vanished. A turbid glacial lake is present at the snout of the glacier, receiving large quantities of sand and silt. The valley in which the glacier now sits is a classic glacial 'U' shaped valley, with steep valley sides and a broad flat valley floor. Today's glacier occupies less than half the width of the valley in which it sits, indicating that it is significantly smaller than its former maximum size, present during older and colder climatic conditions. Figures 6.9 and 6.10 display a series of small north-facing glaciers, largely confined to upper mountain corries/accumulation zones, and small valley glacier regions. These glaciers formed due to the rapid uplift of the Pangong Mountains in which they sit, genetically linked to the Main Karakoram Fault. This mountain location, adjacent to the >4,000 m Pangong Plain, bitterly cold for much of the year, and the north-facing aspect of the glaciers, creates conducive environments for ice accumulations and small-glacier development.

The Earth has witnessed a number of major ice ages during its geological history. Previous ice ages occurred at 2,900–2,780, 2,400–2,100, 715–550, 450–420, 360–260, and 34–0 million years ago. These six major ice ages lasted between c. 300 and 30 million years, respectively (Petterson, 2021). Earth's current ice age began 34 Ma, and was, in part, the result of the formation of the Himalayas, which produced large volumes of greenhouse gas, locking sediments and cooling the Earth from a hotter ice-free world to a world allowing ice to accumulate (Petterson, 2021). Once Antarctica was isolated, surrounded by a cold-water current at 34 Ma, an ice sheet developed. Polar, high-latitude, and high mountain regions are most prone to ice accumulation. The modern-day ice age accelerated from the Quaternary Period, some 2 Ma. Large continental ice sheets grew, particularly in the northern hemisphere, the Himalaya-Tibet region (sometimes termed '*the third pole*') and Antarctica. Ice ages experience periods of ice advances and cold periods (*stadials*) alongside warmer (*interstadial*) glacial-recession periods. Humanoids have been present throughout the Quaternary-Recent Ice Age, with Sapiens appearing 2–300,000 years ago and adapting to numerous cold periods. The Himalayas and Tibet have hosted Sapiens for around 40,000 years: Himalayan Sapiens have experienced several colder and warmer spells during this period (Zhang et al., 2018). Glacial stadials in the Ladakh region have been identified at 500–430,000 years ago, 200–160,000 years ago, 110–60,000 years ago, 60–35,000 years ago, and 12,000–9,000 years ago (Soheb et al., 2022). Ladakhi glaciation styles vary significantly from valley to valley and sub-region to sub-region. Local climates are important.

The arid nature of Ladakh, with annual rainfalls varying between 20 and 145 mm, tends to ice-starve glaciers, except during periods of higher insolation (sunshine) and greater penetration of the monsoon into Ladakh. Anthropogenic global warming stresses glaciers, with glaciers reducing in size globally in recent decades. Inventories of Himalayan glaciers over the past 40 years show a significant or small decrease in glacier size in the south and east Himalayan region (Ahmed, 2025). NW Trans-Himalayan glaciers have, however, mostly retained their ice volumes, or grown slightly in size (Gibson et al., 2017; Soheb et al., 2022). The Baltoro glacial system, within the North Pakistan Karakoram, is particularly extensive. Most glaciers in Ladakh are located between altitudes of c. 6,500 m asl to 4,200 m asl, particularly between 6,000 m and 5,000 m asl. More extensive glaciers occupy higher mountains. Smaller glaciers are mainly located above 5,500 m. Most glaciers face northwards (NE and NW, Figures 6.9 and 6.10), reflecting higher, colder mountain climates on slopes with these aspects. Recent Ladakh studies suggest that Ladakhi glaciers have variably changed over the past 40 years (Soheb et al., 2022). Most Ladakhi glaciers are shrinking by a relatively small amount, except for those in the higher mountain areas, particularly in the NW, with a small percentage growth. Local people express deep concerns about the health of glaciers and almost unanimously articulate change involving glacier shrinkage and lower amounts of glacier-derived water resources (personal communications). All Ladakhi villages rely on water supplies gravity-fed from mountain streams and ice. Some Ladakhis are developing artificial ice reservoirs termed 'ice stupas' or 'ice pyramids'. These structures are produced during cold winter months by dripping water onto a freezing surface, which builds an accumulation of ice of varying sizes, shapes, and dimensions (Figure 6.11). Ice stupas are used up during summer months for water supplies and rebuilt during winter months: they supplement the water supply but could be scaled up. Ladakh now receives hundreds of thousands of tourists, which significantly adds to local water demand.

Glacial landscapes are shaped by erosive and depositional activities of ice. 'U' shaped valleys are one of the most classic glaciated mountain features. Ice is a formidable erosive force, with similar mechanical properties to solid rock in terms of strength and solidity. The dynamic motion of ice erodes rocks through a number of processes such as *freeze-thaw, plucking, and abrasion*. Freeze-thaw refers to the repeated cycles of water freezing and thawing. When water freezes, it expands by 9% and then contracts again as it thaws. The regular freeze-thaw, expand-contract cycles prise rock masses apart from micro to large macro scales. Plucking refers to the freezing of water within rock fractures, which are then plucked and pulled by moving ice. Abrasion refers to the physical force of rocks as they move over the ice-rock interface, tearing, scratching, and shearing the rocks over which they move. Once plucked, rock fragments are used as erosive tools, scratching and drilling into the rocky substrate. As glaciers move downwards, they often leave grooved rock surfaces behind (Marshak, 2019, chapter 22).

Figures 6.12 presents landscapes strongly shaped by ice. Figure 6.12A, from Shinku La, illustrates a higher valley (this view looks down the higher valley towards the deeper valley beyond), tributary to a

FIGURE 6.11 Ice stupa, north of Tanglang La. Note: Ice stupas, or ice pyramids, are human-constructed ice accumulations created during the cold winter months and used to supplement local water supplies during the summer.

FIGURE 6.12 Glaciated 'U' shaped valleys, with steep valley sides and wide valley floors. *Source*: Author's photographs. Note: A and D, Shinku La area, B and C, Rangdum. See text for details.

larger valley in the distance. The higher valley does not reach the larger valley and is left perched high above the larger valley: this type of glacial valley is termed a *hanging valley*. The valley itself has steep valley sides, with sheer rock cliffs exposed. This valley floor is wide, but not flat: rather, it is undulating, because of the presence of glacial deposits (moraines) forming hummocky ground. The river within this valley is disproportionately small compared to the large size of the valley and is termed a '*misfit river*' as it has not been the main erosive agent responsible for the valley. Figure 6.12B, close to Rangdum, illustrates a long valley, with steep valley sides, a wide valley plain hosting extensive glacial deposits, and a misfit river. Numerous side valleys feed into the main valley. Figure 6.12C, also from the Rangdum area, has a high dark ridge in the valley floor composed of recent glacial deposits, forming *a terminal moraine* to a modern glacier that has retreated further up the valley. The ridges on the left-hand side are sharp and jagged, resulting from ice plucking and freeze-thaw processes: sharp ridge crests such as this are termed *arêtes*. Figure 6.12D displays the classic 'U' shape of glacial valleys.

Figure 6.13 illustrates a large 'U' shaped valley through which the River Stod flows in Zanskar. The valley is a more open 'U' shape and has had its valley deposits modified through alluvial processes and human cultivation. Valleys like this form the backbone of agriculture within Ladakh, as flat land, with a climate not too hostile to cultivation, is an imperative for the local mountain peoples' survival, particularly during the bulk of Ladakhi human history, when people were remarkably self-contained.

Glacial landscapes are sculpted not only by erosion but also through deposition. Sections above describe and illustrate how glaciers produce prolific volumes of boulders, blocks, gravels, sands, and mud. These glacial clasts are eventually deposited and form various deposit shapes and landforms themselves. A general term for glacial deposits is *till*. Lithified forms of till found in the geological record from ancient glaciations are termed *tillite*. The simplest form of till is produced by rapid deposition of material from ice, producing very poorly sorted deposits which are often referred to as *boulder clay*, which neatly describes the extreme variation in clast size sorting, from boulders to clay/mud. Even these simplest deposits can be shaped by moving ice, moving them to the sides of glaciers, creating linear ridges of till at valley margins, termed *lateral moraines*, or pushing sediment into the upper-middle sections of a glacier, producing *medial moraines*. Moraines are visible in numerous figures above (Figures 6.5, 6.6, 6.8, 6.9, 6.10, 6.12). Particularly voluminous till deposits occur at the snout of glaciers and are termed *terminal moraines*. As glaciers recede, they deposit a series of arcuate-shaped till ridges that mimic the shape of

FIGURE 6.13 Cultivated 'U' shaped valley, Stod Valley, Zanskar. *Source*: Authors' photograph. Note: The broad valley floor with acres of cultivation and steep valley sides.

the glacial snout. Advances of glaciers plough through previously deposited terminal moraines, reshaping them. Alongside ice, glaciers are full of water. Water lubricates the ice movement, assists erosion, and carves subglacial tunnels. Till can be deposited largely by water-borne (glacio-fluvial) processes: in glacial contexts, the deposits are termed *'glacio-fluvial-deposits'*. Glacio-fluvial deposits are much better sorted than glacial-only deposits, forming ridges and mounds of gravels and sands. Ripple and current structures are common in glacio-fluvial deposits, such as *eskers* (linear, sinuous ridges of predominantly sand and gravel). Some deposits are 'hybrid' and display both glacial and glacio-fluvial features. See Marshak, 2019, chapter 22, for more details.

Lakes and standing bodies of water are common within glacial landscapes (Figure 6.14). Well-sorted, parallel-bedded to laminated muds form within glacial lakes (termed *varves*), along with turbidites. Within a more extensive blanket cover of till, at the peripheries of glaciers or land once overlain by extensive ice sheets, blocks of ice buried within deposits may take many years to melt. Once they melt, they cause the surface to subside, producing *kettle hole hollows*. An abundance of glacial mud can lead to poor drainage and wetland development.

Figure 6.14 displays environments at the termini of a number of glaciers, Shinku-La, Zanskar. The terrain is devoid of plants within this harsh, high, cold, mountain setting. Lower land is dominated by hummocky, uneven deposits of blocks, gravels, sands, and mud. Standing bodies of water and larger lakes are present. Bedrock protrudes through glacial deposits locally, shaped and moulded by moving ice. Upper slopes contain a series of glaciers of varying sizes. The colour of the ice becomes darker and dirtier towards the snout of glaciers. The lower left photograph in Figure 6.14 presents a close-up of a series of terminal moraine deposits, which can be tracked back to a modern glacier, now higher up the mountainside. The lower right photograph (Figure 6.14) illustrates a glacial lake, one of a number of bodies of standing water that form within a poorly drained landscape. The lower left photograph of Figure 6.14 displays voluminous lenticular and blanket deposits of glacial till recording the former greater extent of glaciers. The upper Figure 6.14 photograph provides a photomosaic panorama of Shinku-La. The sheer scale of grey moraine deposits is apparent. Terrains are varied with few flat areas. Deposits are very poorly sorted. All shapes and sizes of rock fragments derived from glacial erosion are visible, from large blocks and boulders to volumes of gravel and sand. These landscapes evidence glacial history and how time is a significant factor in understanding landscapes. The area was once extensively actively glaciated, as recorded within ground deposits, with ice now only occupying small areas: numerous glacial advances and retreats have probably occurred in this area.

Figure 6.15 presents different aspects of a large zone of hummocky ground, east of Karsha, Zanskar. Glacial-formed depositional ridges vary in shape from being half-hemispherical to linear to more amorphous.

FIGURE 6.14 Panoramic view of glacial depositional landscapes around Shinku La, Zanskar. *Source*: Author's photographs.

FIGURE 6.15 Hummocky ground formed of glacial till, with right-hand side photographs presenting the very poorly sorted nature of these deposits. *Source*: Author's photographs. Note: Karsha area. Zanskar. Upper right photograph contains 25cm ruler length. Lower right photograph contains the ruler extended to its full 1m length.

Features are up to 5–8 m high and can extend for tens of metres. The rounded nature of glacial till features suggests deposition from the base or the edge of a glacier/small ice sheet, and subsequent moulding of the till by the motion of the ice. Right-hand side photographs in Figure 6.15 indicate that glacial features are formed from a wide range of clasts of variable size, as does an inspection of the rock fragments that carpet the surrounding terrain. The largest boulder measures 1 m wide by 3.5 m long. Smallest clasts comprise mud and fine sand. A close clast examination gives an indication of the source region of glacial erosion zones. Clasts are from a variety of rock types, mostly found within the Zanskar-Padum valley region. These glacial features are old features in relative terms, possibly from one of the older glacial periods. Reworking of the valley by rivers and alluvial fans has also reworked the till, with older till preserved in relatively small areas on the valley floor. Recent engineering activities, road building, and construction have also destroyed numerous glacial features, as they provide an easily accessible resource of sand and gravel.

Figure 6.16 presents a road outcrop through a glacial till deposit in the Taru Valley, west of Leh. The left-hand side photograph gives an impression of the whole outcrop, whilst the right-hand side one illustrates a close-up of the deposit. The term 'boulder-clay' is a good description of this deposit, which is very poorly sorted. Largest clasts measure 60–120 cm long; some zones have closely packed large clasts. Other layers are more sorted and sand dominated, suggestive of glacio-fluvial processes. Clasts are rounded to sub-angular. The right-hand side photograph presents a wide range of larger clast sizes and shapes, measuring up to 60 cm. An abundance of yellow sand, silt, and mud is apparent, forming the matrix of the deposit. All clasts are locally derived, within Taru itself.

Figure 6.17 presents images of different aspects of terminal moraines. Figure 6.17 A and C are from the Indus Valley, south of Taru, west of Leh. Figure 6.17 B is from the Karsha area of Zanskar. All features form steep and laterally extensive ridges, marking the edge of a more expansive area of till formed by glaciers and/or small ice sheets. These are probably older glacial deposits, of which only relics remain after reworking by river and alluvial fan (and human) activities. Deposits are poorly sorted with the largest rock fragments up to a metre or so in length. The bulk of the larger clasts are pebble-sized. Sand, gravel, and mud form the bulk of these deposits. Bedding is not obvious, except locally, suggestive of ice-glacial origin rather than glacio-fluvial. These features are now extensively used by people for a variety of purposes. They are quarried for sand, gravel, and larger rock fragments used in road building and general construction. Road building has accelerated recently, with four or even five roads connecting Ladakh with other parts of India. This has led to a greatly increased number of tourists and visitors, which, in turn, has led to a high demand for accommodation, hotels, restaurants, and the hospitality industry. These recent trends have fuelled a surge in construction. Business units are also starting up, requiring materials. Higher

FIGURE 6.16 Glacial till deposits. *Source*: Author's photographs. Note: The left-hand side presents the general character of a large exposure within very poorly sorted till deposits. Note yellow field note book measuring 19 x 12cm. The right-hand side provides a close-up of this boulder clay deposit. Note the 1m ruler for scale. Text above for details. Taru, west of Leh.

FIGURE 6.17 Terminal Moraine features and deposits, Indus Valley and Karsha, Zanskar. *Source*: Author's photographs.

terraces formed by glacial and alluvial fan deposits are selected for settlement and business parks. Large numbers of military personnel require sites for accommodation and often choose the till and alluvial fan-terrace deposits to site encampments.

Figure 6.18 presents an esker feature from Karsha, Zanskar, extending for 1 km in length. Another esker, close to Ganglas, north of Leh, extends for over 5 km. Eskers are linear-sinuous features, composed largely of sand and gravel, often with prominent crest features marking the top of the ridge (Figure 6.18). Eskers form as glacio-fluvial constructs, originating as sub-glacial river and stream deposits. The Karsha esker is now an isolated feature, recording more extensive older glacial activity. As prominent local landforms, they are commonly chosen for the sites of stupas and other sacred monuments. The Karsha esker's crest line is clearly visible, varying in height. The ridge has a wide apron of sediment stretching out several hundred metres on either side of the crest. Whilst the bulk of the esker is formed from reasonably well-sorted sands and gravels, larger fragments of rock are present. The esker is aligned east-west; land is now used for grazing with sheep and goats (often present on the esker).

Figure 6.19 completes our examination of glacial deposits. This deposit is quite different from the ones we have examined above. The photograph is from a quarry within mud and silt in the Indus Valley, west of Leh. This deposit appears to be thickly bedded, but is also formed from thin to very thin beds, on closer observation. Parallel bedding is predominant. Deposits such as these were formed within lakes during glacial periods, and have been infilled with sediment to the point where the lakes no longer exist: some lakes are still present within the Indus Valley and elsewhere, genetically linked to current or past glacial activity. Lake deposits are mostly extremely well sorted and comprise fine-grained sediments.

FIGURE 6.18 Ladakhi Esker features from the Karsha area. Zanskar. *Source*: Author's photograph. Note: Its linear form and sand-gravel character.

FIGURE 6.19 Paleo-lake mud and silt deposits, Indus Valley, west of Leh. *Source*: Author's photograph. Outcrop is c. 6m high.

Lakes are (mostly) quiet, low-energy sedimentological environments. Lakes slowly fill with mud and silt that 'rain down' within quiet waters, forming regular *varve* deposits. Varve deposits are cyclical ones, with thicker and thinner beds/laminations. Thicker layers form during times of greater sediment accumulation (e.g., during summer seasons) and thinner layers form during sediment-depleted periods (e.g., winter seasons). These deposits are extensively researched as palaeo-lakes provide key data about past environments (Sharma and Chaudhri, 2021).

We end the section on ice and landscapes with Figures 6.20 and 6.21 illustrating extensive modern wetlands that form within 'U' shaped valleys and poorly drained glacial (and glacio-fluvial/fluvial) deposits across a broad valley floor. The wetlands are a rare and vital ecosystem within Ladakh, hosting a variety of ecosystem-specific flora and many visiting migratory birds. Insensitive development in areas such as these can drain these unique ecosystems and harm beautiful geological landscapes.

FIGURE 6.20 Wetland environments formed within poorly drained glacial deposits, Rangdum. Source: Author's photograph.

FIGURE 6.21 Extensive wetlands within the valley plain below the dramatic Zanskar mountains, Rangdum Gompa in the lower distance. *Source*: Author's photograph.

REFERENCES

Ahmed, R. (2025). Decoding the fate of Himalayan glaciers under climate change: Impacts, challenges, research gaps, and policy pathways. *Evolving Earth*, 3, Article 100082.

Gibson, M. J., Glasser, N. F., Quincey, D. J., Rowan, A. V., & Irvine-Fynn, T. D. (2017). Recent wetting and glacier expansion in the Northwest Himalaya and Karakoram. *Scientific Reports*, 7(1), 6139. https://doi.org/10.1038/s41598-017-06388-5.

Herren, E. (1987). Zanskar shear zone: Northeast-southwest extension within the Higher Himalayas (Ladakh, India). *Geology*, 15(5), 409–413.

Marshak, S. (2019). *Earth: Portrait of a Planet* (6th ed.). New York: W.W. Norton & Company.

Petterson, M. G. (2021, September 28). Climate explained: What is an ice age and how often do they happen? *The Conversation*. https://theconversation.com/climate-explained-what-is-an-ice-age-and-how-often-do-they-happen-166652.

Searle, M. P., Weinberg, R. F., & Dunlap, W. J. (1998). Transpressional tectonics along the Karakoram fault zone, northern Ladakh: Constraints on Tibetan extrusion. *Geological Society, London, Special Publications*, 135(1), 307–326.

Sharma, V., & Chaudhri, A. R. (2021). Geological observations from a palaeolake basin, Lamayuru, Ladakh, Northwestern Himalaya. *Open Journal of Geology*, 11(6), 175–182. https://doi.org/10.4236/ojg.2021.116010.

Soheb, M., Ramanathan, A., Bhardwaj, A., Coleman, M., Spagnolo, M., Rea, B. R., Singh, S., & Sam, L. (2022). Multitemporal glacier inventory revealing four decades of glacier changes in the Ladakh region. *Earth System Science Data*, 14(9), 4171–4185.

Zhang, X. L., et al. (2018). The earliest human occupation of the high-altitude Tibetan Plateau 40 thousand to 30 thousand years ago. *Science*, 362(6418), 1049–1051.

Ladakhi Fluvial and Aeolian Erosion, Deposition, and Landscapes

7

TALUS FANS, ALLUVIAL FANS, AND LANDSCAPES

This chapter presents an illustrated account of fluvial and aeolian activities within Ladakh. The reader is presented with a range of examples of alluvial fans and river/wind landscapes, systems, and deposits. Examples are used to highlight key features of these phenomena and related natural scientific processes. Although Ladakh is a montane desert, it contains a number of permanent rivers, the Indus being the prime example, alongside rivers that may often be dry or run at low water volumes. Arid environments produce a myriad of wind-sculpted terrains and associated deposits. The chapter begins with alluvial fans, then moves onto fluvial systems and landscapes, and ends with aeolian processes. For further details, readers are directed to Marshak, 2019, chapters 17 and 21, Srivasta et al., 1979, Nag and Phartiyal, 2015, Koul et al., 2016, Kumar et al., 2016, Chahal et al., 2019, Dhir et al., 2020, Phartiyal, 2022.

Figure 7.1 highlights the shape of five alluvial fans present in Padum Valley, Zanskar, within an area measuring 10 km by 10 km. Fans here are 1–5 km long and up to 4 km wide, coalescing and overlapping each other locally (e.g., Tungri alluvial fans). This relatively small part of Ladakh demonstrates how active alluvial fans are within Ladakhi Himalayan mountainscapes. Rivers within the Padum Valley, in turn, rework and reshape alluvial fans: Tungri and Karsha fan fronts in Figure 7.1 have been flattened, with material lost to active alluvial systems in the river valley.

Alluvial fan and related talus fan and sheet formation are key processes for transforming mountain landscapes. They modify the shapes of mountains themselves, create flatter land at the front of mountains, and modify the shape and nature of mountain plains and valleys. Large volumes of material are transferred from higher mountain slopes, cliffs, and hinterlands to the plains/lower land. Alluvial fans often rework and reshape previously deposited glacial material and structures.

Gravity is a major force in the production of talus and alluvial fans. Even in the absence of water, loose material moves downslope under the force of gravity. Loose rock fragments prise themselves from the main mountain mass through the physical effects of diurnal temperature variations and freeze-thaw weathering process. As rock fragments, of various shapes and sizes, fall downslope, they accumulate as *scree or talus* aprons. Talus and scree aprons are reworked by fluvial activity in times of heavy, intense rainfall. Earthquakes result in further downslope movement of talus and creation of additional rock fragments. Gravity always encourages slopes to find the optimal stability slope angle. Under Earth's gravitational field, *angles of repose* (steepest slopes for granular material to exist without sliding) for talus and scree slopes are in the range of 34°–36°, whilst rock-dominated mountain slopes have angles of repose of

DOI: 10.1201/9781003794868-7

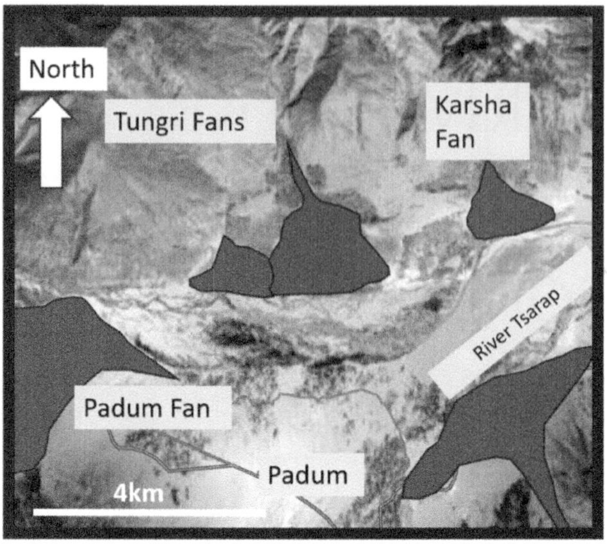

FIGURE 7.1 Alluvial fans (in dark blue) within a Ladakhi landscape around Padum. Zanskar. *Source*: Base Map Acknowledgements. Google.(n.d.). Google-Padum-Ladakh). Retrieved 25 February 2025 from https://www.google.com/maps/search/ place/Padum+194302/@33.4975076,76.8429624,20470m/

FIGURE 7.2 Key elements of an alluvial fan. Nubra. *Source*: Author's interpreted photograph.

c. 30°. Slope angles <<30° maintain slope stability for longer periods of time (Beakawi et al., 2018). The interplay between gravitational force and mountain slopes tends to find an angle of repose under which the slope is stable within Earth's gravitational force field. In practice, equilibrium is rarely reached as our planet is so dynamic. Ancient landscapes within arid terrains, at the centre of continents, far away from active tectonics, approximate the ideal slope equilibrium scenario.

Figure 7.2 is an alluvial fan from Nubra, presenting a near-ideal arcuate shape. Annotations in Figure 7.2 highlight key components of alluvial fans. Narrow valleys within steep mountains transport large amounts of sediment over time, from mountain interiors to the plain/mountain front. As mountain rivers meet the plain, they lose hydrodynamic energy and deposit material. Over time, river channels radiate over a half-circle in front of the narrow river valley exit point. Larger clasts are deposited close to the mountain front (proximal location) with finer-grained material transported to more distal settings. Parts of the fan are active with streams running through and over them, progressively growing the fan. Over time, channels move around the fan. Previously active channels may subsequently become inactive, as fluvial systems move to other locations. Well-formed conical-symmetrical shapes of fans indicate concurrent 'live' fluvial activity within the fan. Conditions are impermanent. Active alluvial fans can become 'dead' or inactive fans. For example, if the river that created the fan is re-routed to another part of the mountain interior, or the local or global climate changes, becoming increasingly arid, conditions for the maintenance of an active fan are removed. Main-valley rivers may swell and erode fans during periods of higher precipitation, or tectonics may change wholesale macro-scenarios and landscapes, through active large scale faulting activity, or causing large scale uplift/subsidence to occur. These examples of changing scenarios testify the changing dynamics of active mountain systems and their potential impacts on alluvial fans.

Alluvial fans, talus cones, and sheets come in a variety of sizes, shapes, and scales. Fan shape is dependent upon a range of conditions: (1) the relative importance of movement of rock clasts by gravity as opposed to fluvial activity. The greater the importance of gravitational forces, the steeper and less extensive the fan. The greater the importance of fluvial processes, the more widespread the fan and the shallower the fan slopes; (2) the more restricted the valley floor, and the closer the valley sides, the less extensive the fan; and (3) the wetter the climate and the more extensive the valley floor, the more widespread the fans.

Figure 7.3 presents an example of a large-scale talus fan from Zangla, Zanskar. This fan is dominated by steep talus slopes, with clasts largely composed of silt, sand, pebble, and cobble-sized lithics. Within this Zangla talus sheet, clasts are largely produced through physical weathering. Freeze-thaw and temperature variation expansion-contraction processes break up mountain rock exposed to the atmosphere, with resultant clasts falling down-slope through gravity.

Figure 7.4 presents an alluvial fan with a steep morphology and limited lateral extent across the narrow valley floor, at the base of the mountains. The fan is dominated by proximal sediments: i.e.,

FIGURE 7.3 Large-scale talus fan. Zangla, Zanskar. *Source*: Author's photograph.

FIGURE 7.4 Steep alluvial fan with limited lateral extent. Zanskar Gorge. Zanskar. *Source*: Author's photograph.

sediments that have not moved far from their source region. Boulders, cobbles, and large pebbles/cobbles are common, with a matrix of finer sands and silts. Gravity is dominant in this fan, but unlike the talus cone presented in Figure 7.3, fan material has spread out to some extent on the valley floor. The ultimate sediment source region for the fan presented in Figure 7.4 lies within the local high-mountain rock exposures/terrain. Clasts are distributed in a fan shape across mountain slopes and the valley floor. This valley floor is constricted within a river gorge, which, in turn, constrains the ability of the alluvial fan to grow too much laterally, resulting in a steep alluvial fan morphology. Boulders have rolled all the way down to the edge of the fan. The deposit here is mixed and shows abundant evidence of water channels and sorting of sediment. Figure 7.4 clearly illustrates how large volumes of material are moved from mountain to valley, modifying valley morphology. Alluvial fans are relatively young and unstable features, constantly changing. Fans themselves erode and rework any sediments that have pre-existed within the area of the fan, such as older glacial deposits. Alluvial fans themselves are, in turn, eroded, reworked, and reshaped by active modern fluvial/river systems.

Figure 7.5 presents an active alluvial fan from Nubra. This fan is an example of a laterally extensive fan which has spread out across the valley floor, at the foot of the mountain. The geometry of the fan is different from those in Figures 7.3 and 7.4, reflecting the greater influence of alluvial processes and the presence of wide, extensive valley floors adjacent to the mountain fronts. This fan has a low fan height/fan radius ratio, with a predominant mountain valley source region. Sediment character varies depending in part on its proximity to the source region. Proximal deposits, close to the source, are coarser-grained and more poorly sorted, whilst distal deposits, for example, close to the edge of the fans, are finer-grained and better sorted. This fan has a number of active fluvial channels, forming an anastomosing network along the upper fan surface, but also penetrates the fan itself, forming intra-fan subterranean channels. One deep channel on the left of the fan has been artificially channelised for water resource and flood control purposes. Humans have excavated parts of the fan on the lower right of Figure 7.5. One village is located close to the exit point of the mountain stream, with some greenery present. Other settlements have developed amongst the more distal parts of the fan. Fans that are less steep and more laterally extensive are often populated and farmed, providing a habitable environment within the harsh, cold Ladakhi mountains. Water supplies are present through active surface channels and from abstraction by drilling into the fan itself. Active fans are *aquifers* (or sources of subterranean water). The fan terminus region is itself being reworked by the river present in the lower main valley. Figure 7.5 demonstrates the dynamic and human nature of alluvial fans in mountain landscapes.

Figure 7.6 presents the virtual experience of standing on top of a number of alluvial fans in the Indus Valley. The broad cobble and boulder-strewn landscape here steeps quite steeply from the mountain front to the River Indus, comprised of numerous coalescing alluvial fans. Figure 7.7 illustrates a section through the same fan. The top and sides of the fans here contain very poorly sorted sediments with clast sizes varying from boulder and cobble to sand and silt. Clast shapes mostly vary between subrounded and subangular. Bedding is visible in Figure 7.7, recording fluvial processes within fans.

Figures 7.8, 7.9 and 7.10 present alluvial fan sediments from the Zanskar Gorge and the Indus Valley. These are examples of coarse-grained, less well-sorted deposits. Largest clasts are cobble-pebble sized, with a sand-silt dominated matrix. Figure 7.8 presents a poorly unbedded unit with angular lithic clasts up to >20 cm in length. The long axes of larger clasts display a lack of orientation or clast imbrication. Matrix is mud-silt grade, mostly massive with local bedding present. This unit was deposited by a high-clast

FIGURE 7.5 Laterally extensive active and cultivated alluvial fan, Nubra. *Source*: Author's photograph.

FIGURE 7.6 Top of the extensive Indus Valley alluvial fans, sloping from the mountain front to the River Indus. *Source*: Author's photograph.

FIGURE 7.7 Section through a distal Indus Valley alluvial fan. *Source*: Author's photograph. This outcrop is around 1.5m high. Note the metre ruler for scale.

FIGURE 7.8 Poorly sorted alluvial fan deposit with abundant angular clasts. Upper fan surface. Zanskar Gorge. *Source*: Author's photograph. Field notebook measures 19 x 12cm.

FIGURE 7.9 Section through a bedded and poorly sorted alluvial fan deposit, with coarse-tail reverse grading. Indus Valley. *Source*: Author's photograph. Metre ruler has cm divisions: 32 cm is visible in this photograph.

FIGURE 7.10 Poorly sorted upper surface of an alluvial fan. Abundant angular clasts. Indus valley. *Source*: Author's photograph.Note metre ruler for scale (25cm length visible here).

concentration debris flow. Figure 7.9 is from the upper surface of an Indus alluvial fan. Here, the largest clast is around 10 cm, with most larger clasts angular in shape. The matrix is a sand-silt deposit. Figure 7.10 is a section through an Indus Valley alluvial fan. Largest clasts are 10–20 cm in size. The matrix is predominantly fine-grained. Bedding is apparent, with quite different facies exposed, derived from a range of debris flow regimes. The lowermost unit comprises a high proportion of angular–subrounded, cobble-sized clasts within a sand matrix. Orientation of clast long axes is not random, but neither highly oriented. Some clast long axes dip to the left, suggesting an element of *current imbrication* (clast long axes aligned and tilted at high angles, with clasts 'leaning' on one another), suggestive of fluvial current activity. High-clast concentrations within a depositing debris flow mitigates against free sorting. This particular unit presents a degree of reverse *coarse-tail grading* (grading of large clast population), indicating a significantly high clast concentration within the parent debris flow. The middle bed in Figure 7.10 has quite a different character to the lower bed, with much smaller, large clasts and a higher matrix abundance. This unit is bedded with concentrations/lenses of larger clasts and local weak coarse-tail reverse grading, suggestive of high-clast concentration debris flow activity, strong current activity locally, and a lower energy regime.

Figure 7.11 and 7.12 presents examples of well-bedded alluvial fan deposits. Figure 7.11 presents a section through a distal fan location within the Indus Valley. Bedding is present throughout the deposit here, with a prominent topographic bench manifesting the 3-D character of the bedding and overall fan morphology. Individual beds are visible that extend across the outcrop. The largest clasts are cobble-pebble grade, with a sandy matrix. Figure 7.12 is from the Zanskar Gorge and displays a well-sorted fine-sand

and silt deposit with ripple structures. The delicate lamination-ripple structures are beautifully exposed. One large cobble-sized clast indicates the power of the currents which deposited the silt-dominated sediment load. This deposit is an example of a fluvial-dominant and current-dominant deposit within an alluvial fan. We now leave alluvial fans behind and move on to rivers, river landscapes, and river deposits.

FIGURE 7.11 Section through an Indus Valley alluvial fan with the gross fan morphology visible alongside internal bedding. *Source*: Author's photograph. Outcrop is 1-2m high.

FIGURE 7.12 Laminated and ripple-bedded silts from a Zanskar Gorge alluvial fan. *Source*: Author's photograph. Field Notebook width is 12cm.

LADAKHI RIVERS, RIVER SYSTEMS, DEPOSITS, AND LANDSCAPES

Figures 7.13 and 7.14 provide context for the rivers of Ladakh, which are all tributaries of the Great River Indus, one of the major rivers of the world. The total length of the River Indus is 3,180 km, draining 1.12 million km^2 (Abbasi et al., 2017). The source of the River Indus is within western Tibet, close to Mount Kailas, and subsequently flows along a NW-SE course through Ladakh in India, and onwards to North Pakistan, turning through a right-angled bend heading southwest through the length of Pakistan.

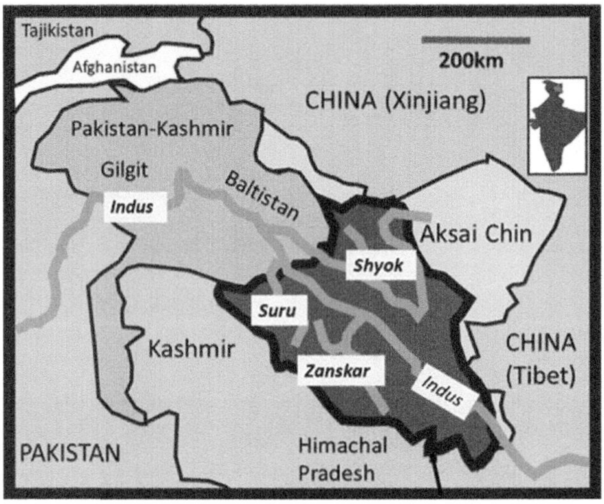

FIGURE 7.13 Main rivers of Ladakh. All rivers in Ladakh are tributaries to the River Indus. Ladakh Union Territory is coloured dark blue and has a thick border line. Aksai Chin and other violet-purple-coloured areas are occupied by China, disputed by India. Pakistan-Kashmir is disputed territory between India and Pakistan but governed by Pakistan. Inset map shows the location of Ladakh within Greater India. North is upwards.

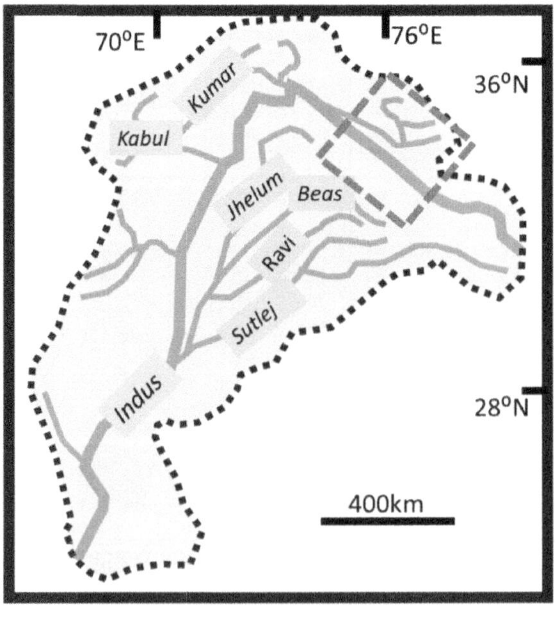

FIGURE 7.14 Map of the River Indus catchment. *Source*: Author's map. Note: The red box indicates the location of Figure 7.13, north is upwards.

The Kabul River system within Afghanistan is part of the whole catchment. The river is of immense importance to hundreds of millions of people throughout the NW Himalayas, the bulk of Pakistan, and the Punjab region of India. The origin of the Indus is intimately linked to the formation of the Himalaya Mountains and the rise of Tibet, particularly strongly controlled by the tectonics of the Ladakh and North Pakistan regions. Sediments, some 45 Ma old, have been discovered within the delta region of the Indus, offshore from Karachi, within the Arabian Sea, suggesting there was a link between Tibet and the present-day Pakistani coast at this point in time. The modern River Indus and its location close to the Indus Suture, and the development of a NW-SE drainage system within Ladakh was set from around 18 Ma (Clift, 2002; Feng et al., 2021).

Figure 7.14 presents the much bigger continental river system context of the Indus, which has an overall triangular shape from a plan view. The catchment system is at its widest point in the north and northeast. The rivers of the Punjab region (Sutlej, Ravi, Beas, and Jhelum) are the biggest tributaries to the Indus, with their source regions in the southern parts of the Himalaya. The Punjab region in both Pakistan and India is one of the most fertile parts of the whole Indo-Gangetic Plain area.

As the River Indus passes through Ladakh, it is in its youthful, smaller stage, with the Zanskar River being as large and important as the Indus at this point. Figure 7.13 shows the main tributaries of the River Indus, namely the Zanskar, Suru, and Shyok river systems. All these rivers are strongly controlled by local tectonics and geological structure, discussed in detail in Chapters 1 and 3. Figure 7.13 shows a rectilinear trellis drainage pattern with all larger river systems orientated NW-SE parallel to the main geological structural trends, and shorter river lengths oriented N-S to NE-SW. The NE-SW sections cut through local faults and geological weaknesses in an imperative to drain ultimately into the Indus, which controls the continental drainage within the NW part of the Indian sub-continent. The Indus flows northwest into Baltistan and Gilgit, where, remarkably, it cuts directly through the rapidly rising Nanga Parbat-Haramosh mountain massif. Uplift of the Karakoram and Ladakh mountains rejuvenates the Indus, strengthening its erosive power immensely.

All rivers begin within upland areas, which collect rainfall from surrounding hills and mountains, and add these to groundwater to initiate a small stream, which may grow into a much larger river. Every river system is surrounded by a *catchment area* that contains all precipitation within the catchment. Each catchment is separated from other catchments via a series of *watersheds,* which decide the direction of flow of falling rain: the rain will enter one catchment or another when it falls upon the watersheds.

Figure 7.15 displays four examples of the upper courses of Ladakhi rivers. Rivers are divided into upper, middle, and lower courses. Whilst these are useful generic terms, all rivers are different. The upper course of rivers is characterised by steep gradients, narrow valleys, and downward erosion. Upper course sediments are very coarse to coarse-grained and poorly sorted. The middle course has gentler gradients, with larger channels that move around wider valleys over time. Middle course rivers erode, transport, and deposit sediments that vary in size from small boulders to sand, silt, and mud. Lower course rivers have wide channels and very wide valley floors: sometimes, valleys are so wide that they are not particularly obvious. Large quantities of sediment, mainly in the form of sand, silt, and mud, are deposited within wide floodplains and ultimately deltas as rivers reach the sea or lake, depending on local geography.

Ladakh is a mountainous region with few flat areas. Ladakh's rivers are mostly upper-course to middle-course rivers. Figure 7.15 presents upper course rivers from (top left, clockwise) Gombo Rangjon, Matho, south of the Indus Valley, Turtuk (Nubra), and Zonghul (Zanskar). The river that runs through the valley below Gombo Rangjon peak feeds into the Tsarap River, which ends up joining the Zanskar River near Padum in Zanskar. The river here occupies a large, glaciated valley with a flat valley floor. The river reworks glacial deposits with a predominant boulder and cobble grade sedimentary bedload, and depositional deposit. Matho and Turtuk rivers occupy more confined 'V' shaped valleys, with limited evidence of significant glaciation, although some local valleys and deposits evidence limited glacial activity. Rivers cut downwards more than laterally within upper courses, leading to relatively narrow valleys. Rivers tend to flow around protruding valley side ridges with a 'wavy' sinusoidal geometry in plan view. Channels are narrow and shallow, and filled with large sediments such as boulders, cobbles, and pebbles. When in flood, the channels move around the valley, reworking and reshaping the valley floor, as well as eroding valley sides. Trees and shrubs have established within river systems, indicating that at least some areas of the valleys are relatively quiet: extensive floods will fell trees from time to time. Matho River joins the

FIGURE 7.15 Views of upper course Ladakhi rivers. *Source*: Author's photographs.

Indus, and the Turtuk River is part of the Shyok River system. Zonghul valley is quite devoid of greenery, with landscapes dominated by rocks and coarse sediments. The river channel is narrow and confined to the lowest parts of the valley, although it will rise in times of flood. The river moves around alluvial fans, reworking the edges of fans and transporting alluvial fan sediment. As with the Gombo Rangjon River, the Zonghul River has begun to meander, gently curving within the valley floor. The Zonghul River joins the Stod River, which, in turn, joins the Zanskar River.

Moving anticlockwise from top left, Figure 7.16 presents river gorges from the Tsarap River, east of Padum, Zanskar, Phuktal, Zanskar, the Zanskar River, below Lingshed, and the Zanskar Gorge close to its confluence with the River Indus. Gorges such as those presented in Figure 7.16 have common characteristics. Valleys are vertical-sub-vertical, with steep to sheer valley sides. They are deep, with tightly constrained valley floors. Channels are powerful and occupy the bulk of the narrow valley floors. Rivers are high-energy systems, especially during times of flood, with high hydrodynamic capabilities, transporting boulders and other large clasts. Rivers can be turbid as they are full of sediment in suspension. They have laminar (fluid, predictable 'straight-line' style water flow) and turbulent flow (more chaotic and irregular flow). Many obstructions within the river bed cause rivers to 'spin' locally, forming whirlpools which can erode the bedload downwards. Rivers erode through a variety of mechanisms, including *dissolution* (dissolving material), *hydraulic action* (the hydrodynamic force of running water), which is greatly amplified by using suspended clasts as 'drilling tools' (*termed abrasion*), and *attrition,* referring to particles and clasts knocking into each other, breaking them up further into smaller particles. Sediment is transported in *solution,* dissolved within its water mass, through *saltation* ('jumping' in parabolic loops along the river bed, down current), in *suspension* (being suspended within the main water body), and through *traction* (being dragged along the channel floor through hydraulic action).

Figure 7.16 presents the most dramatic example of river erosion, with the downcutting of rivers being obviously evident through their vertical valley morphology. Tectonics plays a large part in the erosive

FIGURE 7.16 Ladakhi River Gorges. *Source*: Author's photographs.

power of rivers. As mountains rise, rivers are 'rejuvenated' or energised. Rivers are driven by the imperative to reach a baseline: ultimately the ocean, but locally this may mean a lake or a lower valley. As Ladakh rises through tectonics, it temporarily moves the river further away from its baseline, 'encouraging' the rivers to 'try harder' and 'dig deeper' to get to their desired baseline. The example of the 8,000 m plus Nanga Parbat (a Karakoram Mountain giant) located west of Ladakh, in Pakistan, clearly demonstrates this principle. Nanga Parbat is a young tectonic structure, dating back to around 10 Ma, still rising at around 1 cm/year (Butler, 2019). The River Indus pre-dated Nanga Parbat and was presented with a huge challenge: a statuesque barrier exactly in its way. Rather than moving around the mountain, the River Indus has cut right through the mountain massif.

Ladakhi rivers all feed into the River Indus, which, in turn, seeks its baseline: the Arabian Sea along the Makran coast, close to Karachi, South Pakistan. *Long profiles* of rivers (shape of their trajectory from mountains to ocean) are ideally parabolic in shape. The River Indus has steep gradients through Ladakh and the Karakoram mountains, leading to gentle gradients in the Pakistani lowlands, thus broadly approximating a parabola. However, the tectonics are so dynamic that there are many deviations from a conceptual idealised parabolic long profile. These deviations occur through uplift of mountains and thrust faults, which cause sudden gradient changes called '*Nick Points*' to occur (Marshak 2019, chapter 17).

For half the year, Ladakh is a frozen world with temperatures falling as low as −60°C around Dras in western Ladakh. Temperatures lower than minus 20°C are not unusual, and average daytime temperatures go below zero during much of December, January, and February. During such times, rivers can freeze over for months, forming walking tracks for many in Ladakh. Figure 7.17 presents four images of frozen Ladakhi rivers: the photographs on the left-hand side are of the River Zanskar close to its confluence with the River Indus, and photographs on the right-hand side are from Sankar Valley, close to Leh. The River Zanskar is a larger, fast-flowing river, taking time to freeze over when the cold season sets in. In contrast, the Sankar River is smaller and completely freezes, along with water in soils and fields within the lowlands. The long, cold season constrains many of the alluvial erosive and depositional processes. Flooding

FIGURE 7.17 Frozen Rivers of Ladakh. *Source*: Author's photographs.

is rare or non-existent unless there is a period of thaw, and flowing water is restricted to the deeper parts of channels.

Figure 7.18 presents a view of the Shyok River system in Nubra. Figure 7.19 presents a view of a more valley-constrained river between Khardung La and Nubra. Figure 7.20 presents a photomontage of the Zanskar-Stod-Tsarap River systems in Zanskar, around Padum. These rivers are examples which characterise the middle course of rivers within the context of Ladakh, which has few expansive flat areas, as most of the terrain is mountainous with narrow valleys. Rivers within Zanskar and Nubra's wide valley plains have abundant space in which to move around laterally. River channels are wide and meander across flat valley floors. Rivers are sediment-laden, situated proximally to mountains, that prolifically supply large quantities of boulders, cobbles, pebbles, gravel, sand, silt, and mud. Sediment-rich rivers develop *braiding* morphologies within wide, flatter valley plains. Braiding refers to the way numerous simultaneous channels develop, which meander and weave around one another (as in braided hair or jewellery), separated by mounds or 'islands' of gravel and pebbles, aligned in the direction of flow. Channels move around the whole valley floor over time. If rivers are active, with sufficient water to support them, they restrict the growth of plants and trees, which struggle to permanently take root. Rivers in Figures 7.18/19/20 evidence the dominant role of fluvial systems in shaping river valley floors and the choice of land use. Rivers transport and deposit large volumes of ever-changing, shifting sands, gravels, silts, and coarser materials. It is unwise to develop long-term settlements in such dynamic environments that flood regularly. Much of the valley floor is a floodplain. Shrubs, grasses, and small trees may take root for some time but cannot establish permanent plant communities in any one place. Rivers erode and reshape land features created by preceding glaciers and alluvial fans, moving the scavenged sediment downstream and creating flat alluvial plains, as meanders migrate across valley floors and during times of flood. Rivers cut into their own alluvial deposits as tectonic uplift surrounding areas (Figure 7.19 illustrates this well).

Figure 7.21 evidences the abundant utilisation of alluvial deposits, particularly within the middle river courses. Numerous currently active extraction sites within Ladakhi river channels and flat alluvial floors dig up a range of materials from coarse boulders and cobbles to sands and silt. Extracted materials

FIGURE 7.18 Braided rivers with anastomosing river channels within a wide intramontane valley. Nubra. *Source*: Author's photograph.

FIGURE 7.19 Braided river within an upper glaciated valley, near Khardung La. *Source*: Author's photograph. Note: Note how the river cuts through earlier deposited alluvium, indicating tectonic uplift.

are transported to mineral processing sites, which sort materials into different sizes according to specific construction requirements. These extractive operations provide abundant mineral supplies for local road and bridge construction, alongside village and urban development. Figure 7.22 illustrates how rivers can actually erode deeply into their own previously deposited materials. Figure 7.22 displays deep-river-valley cuts into thick alluvial plain deposits. The level of the plain deposits is visible, forming a flat, extensive intra-montane area of flat land, 20–30 m above current river channel levels. The new valley is steep and strongly 'V' shaped, indicating rapid downward erosion. The former level of the alluvial plain was the baseline that the river previously worked towards for a considerable period of time. At this time, the river meandered widely across the old wide valley floor. Tectonics reactivated uplift through local thrust faulting, which, in turn, elevated the mountains and dramatically increased river erosive power. The tectonic dynamism of Ladakhi river systems within Ladakh is well-documented in Figure 7.22.

FIGURE 7.20 Complex braided river systems roam freely across the widest part of the Padum Valley. Zanskar. *Source*: Author's photomontage.

FIGURE 7.21 Aggregate extraction within the Stod River floodplain, Zanskar. *Source:* Author's photograph.

Figures 7.23 and 7.24 illustrate scenes from the River Indus, west of and close to Leh. The Indus comes to Ladakh from Tibet and forms a wide valley around Leh (Figure 7.24). The river is the lifeblood of Ladakh, bringing water to a desert mountain region. The Indus is the local baseline for all other Ladakhi rivers, into which they ultimately discharge their waters. Figure 7.23, from Phey, west of Leh, is an incised meander bend within the Indus, which locally down-cuts into its own, more ancient, alluvial deposits, leaving its previous higher baseline far above its current level behind. Figure 7.24 is a close-up of Figure 7.25, providing insights into the details of bedded and cross-bedded Indus cobble, pebble, sand, silt, and mud deposits. Individual beds can be tracked across the image in Figure 7.24, with some beds wedging out laterally. Figure 7.25 is from Stakna, east of Leh, and presents an extensive flat, water-rich, valley floor. Alluvial fan deposits slope towards the Indus on both sides of the wide Indus Valley, from the mountain front towards the river itself. This vast sloping valley floor comprises a mixture of alluvial fan, alluvial, and reworked alluvial fan deposits with local lake-mud deposits.

FIGURE 7.22 Rejuvenated mountain rivers cut deeply into previously deposited alluvial sediments, as tectonics elevate the surrounding landscapes, Zanskar. *Source:* Author's photograph.

FIGURE 7.23 Incised meander and eroded older alluvial cliffs. River Indus, west of Leh. *Source*: Author's photograph.

FIGURE 7.24 Close up of Figure 7.23 (west of Leh). *Source*: Author's photograph. Note: Section through an eroded River Indus alluvial cliff reveals beds and cross-beds of pebbles, gravels, sands, silts and muds. The middle distance slope from the mountains to the river marks the upper surface of the coalesced alluvial fan and alluvial deposits.

FIGURE 7.25 Panoramic photomontage of the extensive, wide Indus valley between the Indus Molasse mountains on their right and the Ladakh batholith mountains on the left. *Source*: Author's photograph montage. Note: Valley deposits comprise a mix of alluvial fan, alluvial, and lake mud deposits.

Figure 7.26 presents coarse alluvial deposits from the River Indus. Thick to very thick beds of boulders, cobbles, and pebbles embedded within a sand-silt matrix dominate the upper photograph. Largest clasts are up to half a metre in size, with 5–20 cm long clasts being more common amongst the larger-clast population. Clasts are rounded to sub-rounded, reflecting the impacts of abrasion and attrition processes. Some beds are gravel-dominant with larger clasts mostly around 2–5 cm in length, and a higher proportion of sand-grade particles (lower photograph, Figure 7.26). Ripple, dune, and lenticular bedding testify to the action of high-energy current activity. Only high-energy rivers with large volumes of water, within fast-flowing, turbulent channels, probably under flood conditions, can transport and deposit clasts up to 0.5-1 m in length.

Figures 7.27 and 7.28 display sediments, bedding, and sedimentology contexts of small, upper course rivers from valleys south of the River Indus that drain into the Indus. Upper course river sediments are coarse-grained, dominated by boulders, cobbles, pebbles, and gravels. Figures 7.27 and 7.26 (upper photographs) present images of sediments on the valley floor, close to, and within presently active channels. Largest clasts are up to 30–70 cm, with larger clasts mostly around 8–20 cm. Sub-rounded (with some sub-angular) clasts dominate, typical of rivers proximal to their sources: here, river processes have rounded clast sharp edges, but haven't transported them far enough to mature rounding processes more typical of middle

FIGURE 7.26 Cobble and pebble-rich sand and silt deposits from the Indus gravels, west of Leh. *Source*: Author's photographs. Note: The upper photograph gives an overview of thickly bedded alluvial deposits, with a lower well-bedded pebble and gravel unit and an upper, less well-bedded cobble-rich deposit. The lower photograph illustrates lenticular-bedded sands and cross-bedded pebbles-cobbles indicative of high-energy currents. Upper photograph records a 3m high section, lower photograph measures 60 x 20cm.

FIGURE 7.27 View looking south across the Indus, within a valley located south of the River Indus, exposing the upper levels and a section through alluvial deposits within a smaller desert valley. *Source*: Author's photograph. Note: metre rule for scale.

FIGURE 7.28 Upper photographs illustrate alluvial terrains within small Ladakhi mountain valleys. Source: Author's photographs. Note: Upper left depicts a dry but active river channel with a coarse bedload. Lower left provides a close-up of the top surface of an alluvial deposit, whilst the lower right photograph presents a section through an active river channel. Valleys east of Stok.

course rivers. Figures 7.27 and 7.28 show sections through active channel deposits: these sections are dominated by poorly sorted boulder and cobble-rich sands and gravels with an element of bedding and local clast imbrication. Figures 7.27/28 display river channels within their valley hinterland context. Channels are quite linear in form, shallow, and steep-sided, with coarse channel deposits. For much of the year, the channels are dry but transport high volumes of turbid high-energy water during periods of heavy rainfall. Figure 7.27 (lower left) presents an example of the top surface of alluvium within small Ladakhi river valleys, with subrounded and subangular cobbles and pebbles varying in size between >60 and c. 5 cm, set within a sand-silt matrix. Figure 7.27 (lower right) presents a close-up of a typical alluvial section within these small valley contexts. One large boulder measures 1 m long × 25 cm wide, evidencing occasional very strong flow conditions. The sequence is weakly bedded with lower gravel and sand beds and upper pebble-rich beds.

Figure 7.29 presents thick and long sections of river sands and gravels with subordinate cobble and boulder-rich areas. These sections are well-bedded with strong and laterally extensive bedding-horizontal structures extending across tens of metres of exposure sections, which are 4–10 m high. Beds vary between sand-dominant and gravel-dominated. Some beds wedge out laterally, whilst others are continuous. *Cross bedding* (bedding at an angle to the main horizontal bedding direction) indicates the action of currents. These sections are more typical of middle-course braided rivers, with an abundant sediment load. High-medium energy currents within relatively wide and deep channels transport and deposit sands, gravels, and cobbles.

Figure 7.30 presents finer-grained sediments, varying in grade from gravel, sand, and silt to mud. Laminated (finely bedded) muds and silts form within lakes and floodplains during times of flood. The upper left photograph from the River Indus displays well-formed ripple and dune structures, indicating dominant current flow from right to left. Normal grading with gravels is present in the lower parts of dune structures. The lower left photograph displays laminated muds and silts that locally wrap around a pebble- and gravel-sized clast. Larger clasts and coarser beds evidence variations in energy and water volumes within rivers. Figure 7.30 (upper right) photograph illustrates one modern local (Nubra) context for rippled mud and silt deposition within a low-energy fluvial environment. The lower right photograph illustrates finely laminated lake-mud deposits (below) and fluvial sands and gravels above, indicating how sedimentological facies vary with time, dependent on conditions.

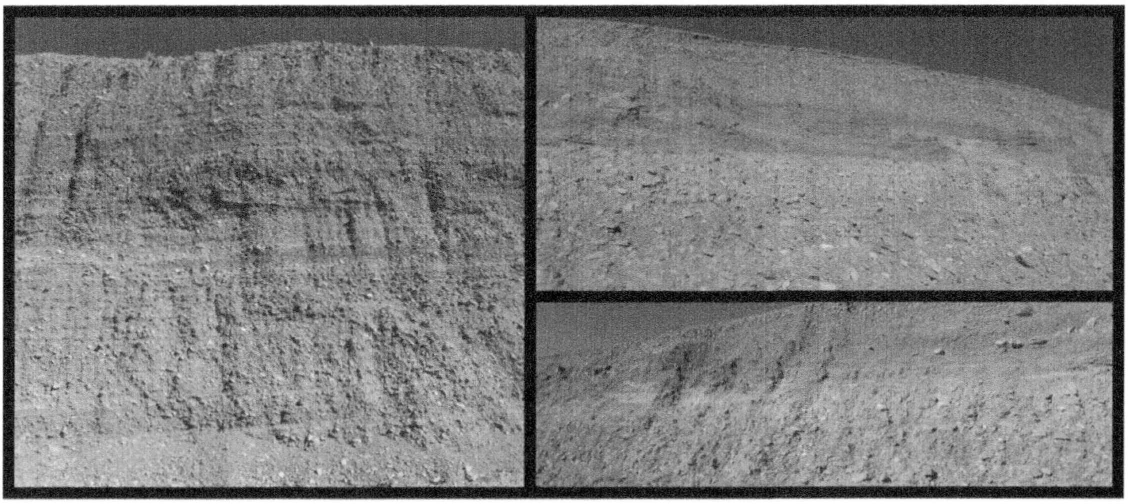

FIGURE 7.29 Large-scale outcrops of river gravel and sand-dominated deposits. Note: The left-hand side photograph is from Phuktal, the upper right from Chazar, near Zangla, and the lower right from Stondey, Zanskar. Sections are (respectively) 10m high (Phuktal) and 4m high (Chazar).

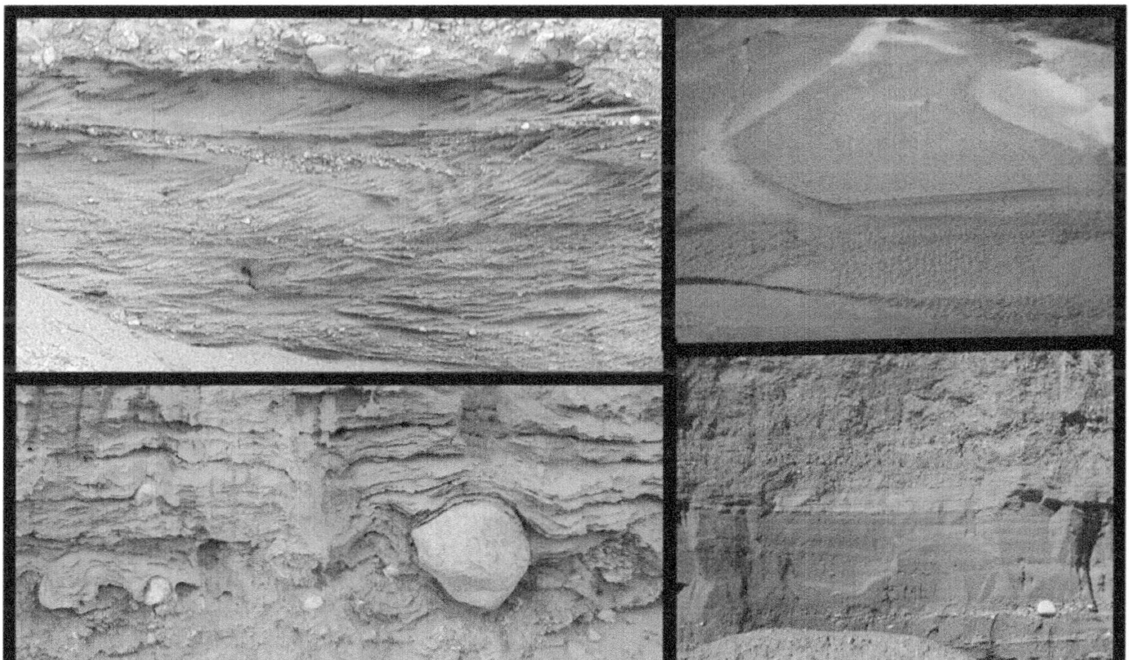

FIGURE 7.30 Sand, silt, and mud deposits from the River Indus (left side), Nubra (top right) and Phuktal, Zanskar (bottom right). *Source*: Author's photographs. Note: Upper left photograph depicts a 30cm high x 60cm long field of view. Lower left photograph measures 35cm long x 15cm high. Lower right photograph depicts a 2m high section. Upper right photograph depicts a 4m x 5m scene.

THE ACTION OF WIND SCULPTING ARID LANDSCAPES OF LADAKH

Ladakh is a mountain desert (Figure 7.31), kept alive through the action of the River Indus and its tributaries. Average annual rainfall is around 80–100 mm, with 300 days per year of mainly arid blue skies. Insolation levels are very high, particularly during the summer months. Flash floods occur from time to time, but these are relatively rare events. Ladakh is within the rain-shadow region north of the main Himalayan watershed. Like many continental interiors, including central Australia, Tibet, Mongolia, Xinjiang, Kazakhstan, and the rain shadow regions of the Andes, and SW USA, Ladakh experiences an arid, sun-drenched environment with high levels of ultraviolet radiation, and only sparse areas of natural vegetation (Figure 7.31).

During the summer months, when the ground is ice-free, Ladakh's sediments can become airborne. During the winter, surface sands, silts, and muds are frozen and bound together with ice. When the ice melts and surface sediments dry out, they are no longer bonded to other particles. Winds can pick surface sediments up and transport them.

Like any fluid current, wind can transport particulate matter (Bagnold, 1941). Air is not as powerful as water. On Earth, winds normally transport particle sizes from mud through to coarse sands within airborne currents, although larger particles can roll across the surface. On Mars, with a much thinner atmosphere, silts, muds, and fine sands only become airborne. As with water currents, particles are transported through traction, saltation, and suspension.

When sediments are airborne, they can become a sand-blasting agent, eroding, shaping, and sculpting rocks, non-cemented surface deposits, mountains, and valley floors. Sand particles become highly rounded and well sorted. Fluted terrains result from sand erosion: the more easily eroded rocks become indented and shaped by wind currents, whilst more resistant rocks stand proud. Valley floors become depleted in finer particles, leaving coarser gravels, cobbles, and boulders behind to form *'desert pavements'*. Rocks can become highly oxidised with the intense levels of insolation and ultraviolet radiation. A skin of *'desert varnish'* comprising oxidised/hydroxidised minerals coloured red, brown, yellow, and ochre penetrates several micrometres or deeper beneath exposed rock surfaces. As with all currents, wind deposits material when its energy diminishes. Most rock outcrops throughout exposed mountainsides are coated with sand and silt. Bowl-topographic structures within mountainsides accumulate large masses of

FIGURE 7.31 Mars or Earth? Hinterland of Phyang, west of Leh. *Source*: Author's photograph. Note: This is an aeolian-dominated landscape, with sand-covered topography and wind-blasted rock outcrops.

sand and silt. Extensive plains which experience limited fluvial and related water activity can accumulate sands and silts, which are then shaped into sand dune fields.

Predominant wind direction decides which slopes are highly wind eroded and which areas of a landscape accumulate sand and silt. Rock faces with an aspect towards the predominant wind direction become extensively sculpted by sand-blast style erosion. Depressions, bases of mountains, and bowl-amphitheatre-like topographic structures whose aspect faces away from predominant wind directions (lee side slopes) accumulate sediment (for more general background, see Marshak, 2019, chapter 21).

Figures 7.31 and 7.32 present typical examples of highly arid landscapes in Ladakh. Remarkably few plants are present within these landscapes. Abundant vegetation is restricted to fluvial environments, along river channels, in areas where the water table is relatively close to the surface, and in irrigated zones close to human settlements. Arid and salt-tolerant grasses, herbs, and shrubs grow throughout Ladakh, together with mosses and lichens. Wildlife is adapted to diets rich in desert grasses, herbs, berries, and shrubs, with predators such as snow leopards preying on wild deer and sheep. Readers are directed to the following references to further explore Ladakh mountain ecological aspects; (Fox, 1994, Schweingruber et al., 2018, Gaur et al., 2024, Giri et al., 2024.)

Arid landscapes of 7.31 and 7.32 could almost pass for Martian terrains. Brown, yellow, red, and ochre colours are the predominant palette. Mountains are covered in varying thicknesses of mud, silt, and sand. Areas within the lee of winds accumulate metres to tens of metres of sand and silt. Rockier outcrops display winnowed and peaked shapes. Rare falls of rain lead to the development of gullies that indent sand deposits. Valley floors are boulder and cobble-rich. Larger clasts stand proud above the valley surface, separated by coarse sand and gravel. Large clast sides can become flattened and sculpted by wind-sand-blasting.

FIGURE 7.32 Arid Ladakh landscapes. *Source*: Author's photographs. Note: From top left, going clockwise: East Leh, Lower-Taru, Upper-Taru, Sankar.

Water is not far away from the images in Figures 7.31 and 7.32, with the River Indus and smaller valley streams bringing water to these waterless landscapes. Groundwater surface levels can be deep: research in areas west of Leh discovered groundwater depths at 65–75 m below ground level, and shallower (c. 25 m below ground level) depths within valley regions close to river channels (Rayeez and Riyaz, 2020).

Figure 7.33 presents (left-hand side) a highly jointed granite outcrop. Wind has accentuated gaps between joints (fractures) within the rock mass, and shaped granite outcrop form, giving it a smooth and rounded tor-like appearance. The right-hand side dusty photograph captures large masses of airborne silt, mud, and fine sand, which are regular occurrences during summer afternoons as thermal convection currents mature in the atmosphere.

The upper left photograph of Figure 7.34 presents a highly fluted sequence of mud deposits around Lamayuru. Wind has winnowed and furrowed the outcrop of this unconsolidated lake mud deposit. The lower left and top right photographs display highly winnowed and fluted alluvium deposits (Chazar area, Zanskar) and weak lime-rich sandstones (Sengge-La, Zanskar). Wind preferentially erodes finer particles within the alluvium and weaker parts of sandstones. Tall towers stand proud of the alluvium. Highly fluted 'corridors' upstand above a more highly eroded sandstone unit. The lower right photograph presents a boulder-rich valley floor from Nubra. Surface deposits are fine-sediment depleted, whilst boulders are wind-etched, smoothed, and sculpted by sand-blasting winds that travel at high velocities down-valley.

Figure 7.35 exhibits elements of wind erosion and deposition. This granite outcrop is significantly fractured. Fractures are wind-blasted and widened through erosion. Within the same image, deposition is evident. Many rock faces are coated with sand and silt particles, whilst sand accumulates in greater thicknesses within areas located within the lee/sheltered sides of rock faces and hollows within outcrops that encourage wind eddies to form and deposit suspended sand and silt.

Figure 7.36 presents a photograph of abundant sand accumulation within a topographic bowl or hollow surrounded by mountains. This amphitheatre-like structure traps sand and protects it from erosion, allowing it to accumulate with thicknesses of over ten metres locally. The topographic setting conditions air currents and motions, with eddy-swirl-like currents forming within the topographic bowl and rapidly depositing suspended particles.

Photographs within Figure 7.37 present a range of examples of sand accumulating within the face of long, extensive ridges. The ridges have developed hollows within mountain faces, deep enough to facilitate the accumulation of significant thicknesses of sand and silt. The deposits are not eroded, although surfaces are reworked by wind, and are persistent features over time.

Figure 7.38 presents dune fields from Nubra. Here, extensive, flat, alluvial plains accumulate significant, but not overly abundant, quantities of sand and silt, shaped by wind into dunes with large lateral bases, relatively small crests, and a concave front elevation: such dunes are termed *'barchan' dunes* and advance

FIGURE 7.33 Granite tor rock shapes, Taru (left-hand side). Summer dust storm, Zanskar (right-hand side). *Source*: Author's photographs.

FIGURE 7.34 Wind-eroded landscapes from (top-left, clockwise) Lamayuru, Sengge-La area, Zanskar, Nubra, Chazar, Zanskar. *Source*: Author's photographs. Note: See text for details.

FIGURE 7.35 Granite outcrop with fractures widened through wind erosion, and protected areas accumulating wind-blown sand. *Source*: Author's photograph. Note: Image presents c. 10m of terrain from top left to bottom right.

FIGURE 7.36 Large topographic amphitheatre bowl at the centre of this image is a significant depositional centre for wind-blown sand. Leh. *Source*: Author's photograph.

FIGURE 7.37 Three panoramic examples of major blown-sand accumulation zones across and within extensive topographic ridges. *Source*: Author's photographs. Note: Upper photograph, Sindhu Ghat near Shey, middle photograph from Saboo, and lower photograph from Indus Valley.

from left to right within the photographs. Less active dunes become vegetated. Barchan dunes generally form on hard unvegetated surfaces, within unidirectional or dominant wind direction conditions. They form in environments where abundant sand is not available: with increasing sand, more contiguous dunes form. Figure 7.38 provides evidence for dominant left-to-right wind directions (Marshak, 2019).

Sediments that accumulate in dunes and sand fields, as depicted in numerous figures above must come from somewhere. Much sediment is derived locally and transported within valleys. Very fine-grained sediment can reach high atmospheric elevations and travel for thousands of kilometres. Figure 7.39 provides clues for the possible ultimate source region of the bulk of sediment presented in previous figures. Desert surfaces become relatively depleted in fine-grained particles and enriched in coarse particles. Abundant pebbles, cobbles, and gravel-sized clasts, up to c. 20 cm, are present at the ground surface

FIGURE 7.38 Barchan dune fields from Nubra. *Source*: Author's photograph.

FIGURE 7.39 Desert surface showing an abundance of larger clasts and a depletion in finer-grained clasts.

of most of Ladakh. Coarse sand and fine gravel constitute around 70–80% of ground surfaces, with the remainder comprised of coarse gravel, pebble, and larger clasts. Wind winnows and sifts continuously at the interface of wind and ground surface. Finer particles can become airborne and transported far from their original position. Coarser particles may 'hop' in saltation parabolic loops in the air or roll across the surface, eventually being trapped by the mountains at the edge of valleys. Figure 7.39 presents a *desert pavement*, a feature common to deserts around the world, from Arizona to Algeria, Namibia to Mongolia, and Yilgarn, Australia, to vast cold desert landscapes of Mars. Larger clasts become strongly sand blasted over protracted periods of time, creating 3 or 4 flattened lithic faces, often producing flattened, pyramid-style, topologies, aligned to predominant wind directions. Such sand blasted and shaped clasts are termed 'dreikanter', German for '3-edges'.

REFERENCES

Abbasi, S. S., Ahmad, B., Ali, M., Anwar, M. Z., Dahri, Z. H., Habib, N., Hussain, A., Iqbal, B., Ishaq, S., Mustafa, N., Naz, R., Virk, Z. T., & Wester, P. (2017). The Indus Basin: A glacier-fed lifeline for Pakistan; HI-AWARE Working Paper 11. Himalayan Adaptation, Water and Resilience (HI-AWARE) HimalDoc. https://lib.icimod.org/records/e29cr-d4s95.

Bagnold, R. A. (1941). *The Physics of Blown Sand and Desert Dunes.* Methuen & Co. Ltd.

Beakawi Al-Hashemi, H. M., & Baghabra Al-Amoudi, O. S. (2018). A review on the angle of repose of granular materials. *Powder Technology*, 330, 397–417.

Butler, R. W. H. (2019). Tectonic evolution of the Himalayan syntaxes: The view from Nanga Parbat. In P. J. Treloar & M. P. Searle (Eds.), *Himalayan Tectonics: A Modern Synthesis.* Geological Society, London, Special Publications, 483, 215–254. https://doi.org/10.1144/SP483.7.

Chahal, P., Kumar, A., Sharma, C. P., Singhal, S., Sundriyal, Y. P., & Srivastava, P. (2019). Late Pleistocene history of aggradation and incision, provenance and channel connectivity of the Zanskar River, NW Himalaya. *Global and Planetary Change*, 178, 110–128. https://doi.org/10.1016/j.gloplacha.2019.04.013.

Clift, P. D. (2002). A brief history of the Indus River. *Geological Society, London, Special Publications*, 195(1), 237–258.

Dhir, R. P., Singhvi, A. K., & Kar, A. (2020). Evolution of arid landscape in India and likely impact of future climate change. *Episodes Journal of International Geoscience*, 43(1), 164–180.

Feng, H., Lu, H., Carrapa, B., Zhang, H., Chen, J., Wang, Y., & Clift, P. D. (2021). Erosion of the Himalaya-Karakoram recorded by Indus Fan deposits since the Oligocene. *Geology*, 49(9), 1126–1131. https://doi.org/10.1130/G48445.1.

Fox, J. L., Nurbu, C., & Chundawat, R. S. (1994). Wildlife conservation and land-use changes in Trans-Himalayan region of Ladakh, India. *Mountain Research and Development*, 14(1), 39–60.

Gaur, M. K., Goyal, R. K., Kanwar, M. S., Singh, N., Choudhary, V., & Chichaghare, A. R. (2024). Prospects and challenges for biodiversity in the anthropogenic-induced cold arid landscape of Ladakh, India. *Japanese Journal of Research*, 5(6), 1–12.

Giri, L., Angmo, S., Gyalpo, P., Chuskit, S., Angmo, N., Rana, S., & Mukherjee, S. (2024). Biodiversity of Leh town of Ladakh UT. G.B. Pant National Institute of Himalayan Environment, Ladakh Regional Centre.

Koul, M. N., Raina, R. K., Bhardwaj, A., & Koul, A. (2016). Research on geomorphic features of Ladakh. *Environment Conservation Journal*, 17(1–2), 119–128.

Kumar, A., Srivastava, P., & Meena, N. K. (2016). Late Pleistocene aeolian activity in the cold desert of Ladakh: A record from sand ramps. *Quaternary International*, 425, 411–424.

Marshak, S. (2019). *Earth: Portrait of a Planet* (6th ed.). New York: W.W. Norton & Company.

Nag, D., & Phartiyal, B. (2015). Quaternary records of the Indus River, Ladakh, NW Himalaya: A review. *Palaeogeography, Palaeoclimatology, Palaeoecology*, 440, 222–234.

Phartiyal, B. (2022). Sedimentation, tectonics and climate in Ladakh, NW Trans-Himalaya: A review. *Geological Society, London, Special Publications*, 523, 31–51.

Rayeez, A. P., & Riyaz, A. M. (2020, December). Evaluation of aquifer system and water level depth in parts of Leh district, Ladakh region, India [Conference presentation]. International Geographical Union (IGU) India, International Conference (Virtual).

Schweingruber, F. H., Steiger, R., & Hellmeier, C. (2018). *Anatomy, Age and Ecology of High Mountain Plants in Ladakh, the Western Himalaya.* Springer International Publishing.

How Mountains and Tectonics Shape Ladakhi People

8

INTRODUCTION

Mountains create illustrious topographic edifices, regarded by some as barriers, others as fortresses, and, for many, places of refuge, adventure, and inner peace. Few humans fail to be moved by the sheer natural forces, prescient in every peak, valley, glacier, and gorge. The great mountains of Earth, regardless of their geological age, are the primary product of tectonics, sculpted by ice, wind, water, and humans. Young mountains, such as those that run the whole length of the west of the Americas, the Alps and Southern Alps of Europe and New Zealand, the volcanic mountains of Indonesia, and the great chain of mountain ranges from the Himalaya and Tibet through Iran and Turkey to the Caucasus and Carpathians, are all tectonic creations: a testament to the dynamism of planet earth (Figure 8.1).

Capturing the majesty of mountains is a challenge for even the most talented of writers and poets. Andrew Harvey first published a seminal book in 1983 (later republished in 1984 and 1993), entitled *A Journey in Ladakh*, which has been widely read, particularly by earlier Western travellers to Ladakh. The book portrays a young man's inner and outer journey, fascinated by Ladakh and the promises it offers to a spiritual seeker. Harvey's writing includes evocative descriptions of landscapes, emphasising the harshness of Ladakh, and the profound silence desert mountain-scapes bring. He discusses the vibrancy of rock formations the changing hues of light at different times of the day, and Ladakh under moonlight. Harvey describes how the colours of cliffs change as light and shadow plays with mountains and valleys. Andrew is constantly surprised at the universal signs of a human presence, even within the most remote of localities. Quotations such as these below epitomise Andrew Harvey's thoughts:

> Nothing I had read or imagined prepared me for the splendour and majesty of the mountains that first day; that was the first gift Ladakh gave me, a silence before that phanstasmagoria of stone, those vast wind-palaces of red and ochre and purple rock, those rock faces the wind and snow had worked over thousands of years into shapes so unexpected and fantastical the eye could hardly believe them.'...
>
> 'That silence! It is the quality of that silence that no words can convey, that silence sustained over millennia, that silence of snow and rock and water in which every sound and movement was contained'.... and....'
>
> Each mountain existed in its own contortion, each crag of purple rock, each landslide of scree, each winding dark stream, each small shrub clawing the sides of the road, each bird seemed to be so full of its own essence that it hovered on the brink of dissolution.'

Harvey makes numerous observations of links between Ladakhi people, culture and religion, and the interconnections between mountains, landscapes and people, each shaping the other. He describes Ladakhi people as joyful, welcoming, resilient, and connected: people who have integrated their traditional

DOI: 10.1201/9781003794868-8

FIGURE 8.1 Wherever you look in Ladakh, stunning tectonic landscapes appear. *Source*: Photograph by Lanka Nanayakkara.

lifestyles in harmony with their mountain environment, in turn creating a human-nature inspired '*sacred geography*', through their construction of monasteries (called gompas in Ladakh), stupas (or chortens), prayer flags, statues, and sculptures throughout Ladakh. Quotations such as these presented below capture his sentiments:

> the signs everywhere, in this wilderness, of human companionship (are present); everywhere along the track there are mantra stones (stones with sacred inscriptions), stones with the letters OM MANI PADME HUM carved on them, breathing their sacred words silently between the rocks. … There are latos, shrines to the spirits of the rocks and the air, with yak horns and bones. There are prayer flags, lonely, ragged but enduring, high on the pinnacles of rock above me, flapping in the wind that communicate the ubiquitous presence of humanity across this vast wild landscape.

HIMALAYAS, TIBET, AND CHANGING NATIONS

Mountains' shape, or at least influence, politics and country boundaries. The Himalayas and Tibet broadly separate the Indo-Aryan peoples of the Indian subcontinent from the Sino-Tibetan-Mongolian-Kazakh peoples of central Asia. Within the Trans-Himalayan region, the nations of Tibet, Bhutan, Nepal, Myanmar, Ladakh, Kashmir, Gilgit-Baltistan-Hunza, Afghanistan, Pakistan, India, and others have all been profoundly shaped by the mountains themselves. A number of in-depth books have been published which examine different aspects of the Himalayan people, landscapes, history, and politics. These include the following: *Himalaya: A Human History* (2021) by Ed Douglas, *High* (2022) by Erica Fatland, *A Cultural History of Ladakh* (2024) by Nawang Tsering Shakspo, *Tenzing, Hero of Everest* (2003) by Ed Douglas, and *The Gilgit Game* (1979) by John Keay. If we take the long view of kingdoms and states that have existed within the broad Trans-Himalaya region, we will dig up names of territories that include:

Nepal, Sikkim, Bhutan, Tibet, Ladakh, Kashmir, Kulu, Garhwal, Kumaon, Lahaul, Mustang, Gandhara, Bactria, Chitral, Hunza, Gilgit, Baltistan, and Nagar. These self-governing, or quasi-autonomous provinces, kingdoms, and countries existed at different times in history, many of which remain as named regions within modern-day Afghanistan, Pakistan, India, and China, alongside independent Bhutan and Nepal. The sheer number of historical political units reflects a diversity of peoples, languages, and cultures, able to exist within mountainous Himalayan geographies, isolated from others, at least to a degree. In turn, this reflects how mountains shape nations, people, and culture. For much of history, many mountain districts remained remote and isolated for hundreds of years, allowing peoples within to develop unique identities, languages, and ways of life. Whilst the bulk of populations remained within restricted geographic areas, traders, travellers, and explorers developed walking routes across mountain passes, connecting communities and nations. Although mountains can form relatively impassable boundaries, there are always 'weaknesses' within the peaks. Cols form lower ridge areas that become mountain passes, at least during the warmer seasons, across which trans-mountain routes are created. The very name Ladakh means 'land of mountain passes', noting their importance and very existence. Mountain passes are mostly linked to geology and structure. Highly fractured rock within fault zones, for example, is more highly eroded than surrounding unfractured ridge rock, thus lowered in altitude, presenting opportunities for humans and animals to cross mountain passes.

PHYSICAL CHALLENGES AND ADAPTATIONS

Mountains present a series of challenges to human settlement, even today, being among the most sparsely populated parts of Earth. Some of the most hostile Earth environments exist within mountain belts, limiting biodiversity (and encouraging endemic adaptation) within wildlife, as well as providing challenges for humans. Climates are more extreme compared to low-lying neighbouring regions. Higher areas of the Himalaya-Tibet region contain valleys and plateaux at altitudes of 3,000–5,000 m asl, with air pressures at 70–50% of sea level air pressure. Temperatures can fall as low as −60 °C (in Dras) and commonly −20 C to −30 °C in Ladakh, with warmest temperatures reaching highs of 33 °C to 37 °C, but mostly 23 °C to 24 °C. Four to six months of the year experience sub-zero average temperatures. The lee or rain shadow zone of mountains brings aridity, with <100–c.250 mm of annual rainfall is experienced across the Ladakh region. Flash floods occur every 3–5 or 10 years across Ladakh. Glacial lake outflow floods pose local hazards, particularly with rising temperatures and consequent increased ablation of glaciers. Flat and cultivatable land is limited within mountains, along with variable moderate-low levels of fertility within relatively thin, immature, and organic-poor soils. Only hardy crops such as barley, pulses, turnips, buckwheat, and potatoes can survive the short, colder growing seasons. Some fruits, such as apricots, are grown in parts of Ladakh. Yaks, dzos, goats, hardier cow and sheep breeds, donkeys, and (locally) camels are reared as Ladakhi farm animals. Human cardiovascular and respiratory adaptations to high altitudes include increased lung capacities, lower haemoglobin blood levels, and more efficient heart and muscle functionalities (Gulia and Singh, 2021). Higher exposure to ultraviolet radiation, due to the presence of a thinner atmosphere, can prematurely age exposed skin and adversely affect vision. Water resources are constrained and require careful, sustainable management. Conversely, mountains act as 'water towers' and become areas of high strategic importance to lowlanders in surrounding regions, who rely on mountain-sourced river water. This leads to geopolitical tensions. Larger powers seek to control strategic and critical mountain hinterlands. Geographic isolation brings benefits and disbenefits. Many mountain people enjoy the isolation and unique cultural character that isolation encourages, alongside self-sufficiency and resilience. On the other hand, access to services and facilities that communities increasingly demand is limited, including schools, universities, hospitals, and employers. For thousands of years, many (most) high mountain areas had walking-only access during the summer months to other regions. The past few decades have witnessed a revolution in mountain road and tunnel building. It is now possible to drive to within a few kilometres of the Tibetan side of Everest base camp and catch a train from Beijing to Lhasa. Mountain airports are more numerous. Visitors to Ladakh have exponentially increased from 35,000 per year in 2005 to over half a million in 2023 (Directorate of Tourism, 2024), within a territory with a permanent population of 270,000–300,000, bringing new challenges to a fragile ecosystem. In summary, mountain and high plateau areas have historically provided natural borders and boundaries to kingdoms and states. In spite of a myriad of physical challenges, hardy humans have settled high mountain regions for millennia, developing unique, resilient cultures. Only the most recent decades have opened up previously isolated areas to many thousands of visitors and increased access for mountain peoples to highly populated lowland neighbouring regions. Political tensions can hone in on mountainous border regions, with recent military disputes between India and China, India and Pakistan, and China's increasing control across Tibet and Xinjiang.

EARLY HUMANITY AND ITS TRACES WITHIN LADAKH

As Bob Marley insightfully and musically declared in his song *Buffalo Soldier* (Marley, Island Records, 1983): *'If you know your history / Then you would know where you're coming from / Then you wouldn't have to ask me / Who do I think I am?'*. He captured the idea that we cannot deeply understand the current state of cultures without knowing their history. Ahtushi Deshpande, in her book *Speaking Stones* (2024), comprehensively photographed, recorded, and explained Ladakhi rock art: a lasting testament to earlier peoples within Ladakh.

Rock art is a vivid, palpable demonstration of links between tectonics, geology, microbiology, and humans. All Earth deserts develop desert-varnish surfaces on exposed rocks within sheltered arid environments. Desert aridity results from a combination of factors but is mostly formed through: (1) great geographical isolation from the ocean, as in the centres of continents, such as Australia, Asia, and Africa; (2) regional atmospheric-climatic conditions that produce cloudless skies for much of the year; and (3) tectonic mountain barriers, orographic rainfall, and hinterland rain-shadow dryness. Desert varnish is a yellow-brown-black-ochre coloured surface 'rock skin', 100–200 μm thick, forming on exposed rock surfaces. Desert varnish is composed primarily of oxides of manganese and iron, together with silica, clay minerals, and microbes. Varnishes form within desert environments sheltered from wind abrasion/sand blasting, and within dry environments, although water plays a role in supporting microbes, and through evaporation of, for example, dew. Physico-chemical and biochemical processes (the latter, particularly within manganese-rich black varnishes) involving oxidation, wetting and drying, drive desert varnish production. Geochemical reactions that occur between airborne sand and clay particles coating rock surfaces, and the rock itself, produce varnish layers at extremely slow rates, taking thousands of years to make even a micro-metre thick deposit. Some researchers debate the importance of microbial life in the production of desert varnish, whilst others prefer non-biological origins. This argument takes on another level of importance with respect to Martian science, as desert varnish exists on Mars: could this be evidence for microbial life on Mars? For further reading, see: Krinsley et al., 1995, Perry et al., 1978, and Chernicoff and Whitney, 2007. Desert varnish surfaces provide a landscape-scale artistic canvas for humans to create images that last for thousands of years.

Ahtushi Deshpande, for the fieldwork relating to her book, *Speaking Stones*, walked through valleys, cameras, lenses, and notebooks in hand, across all regions of Ladakh, over many years. She uncovered archaeological and geological discoveries made by the Geological and Archaeological Survey organisations of India, which helped her to understand the scientific and historical context of rock art (created within desert varnish). Rock artworks, or *petroglyphs*, are made by carving desert varnish with stone and metal tools. Stone (lithic) tools and campsite remains of early peoples have been found in Ladakhi locations such as Changthang, Nubra, Alchi, Nurla, Hemis, and Khaltse, relatively dated at Lower-Middle

Palaeolithic periods, with indicative ages of 700,000 to 300,000 years before present. The oldest known rock art in the world is dated at 45,500 years before present, from Sulawesi, Indonesia. Petroglyphs from Ladakh are thought to extend across a 30,000-year period, spanning Neolithic, Bronze, and Iron Ages. The oldest hand-span impressions may be 30,000 years old. The great bulk of petroglyphs are thought to be aged between 5,000 and 1,000 years before present: petroglyphs linked to modern cultures continue to today (Figure 8.2). Rock art depicts animals, humans, and figures that are part-human and part-animal or mythical. Ibex is the most common animal within rock art, appearing in almost half of all studied petroglyph sites. Yaks and wild sheep appear in 15% of paintings, and the human form similarly in around 15% of all paintings. Other animals depicted are birds, dogs, horses, camels, deer, antelope, and snow leopards. Ritual and hunting scenes are presented, with humans on horseback or accompanied by dogs, often with bows and arrows. Most scenes involve animals only. Styles of art change over time, providing a method of classification and relative age dating. The body shape of humans and ibex, styles of ibex horns and skulls, and other characteristics are used to classify rock art. Studies of rock art reveal different human lifestyles and elements of culture, suggesting periods of nomadic and transhumance ways of living, trading, and cultural links of Ladakhi people with others from Tibet, Kazakhstan, Uzbekistan, Mongolia, North Pakistan, and the western Trans-Himalaya, together with later contacts (from the 8th century) with literate Central Asian societies. A suggested chronology of the past 11,000 years includes: first long-term Ladakhi settlements from 11,000 years before present (Neolithic Period); local Bronze Age dated at 5,000–4,000 years before present; local early Iron Age dated at 11,000–5,000 years before present, and Late Iron Age dated at 6,000–1,200 years before present. The historical period dates back to 2,300 years before present, with Tibetan Empire extensions into Ladakh during the 7th–9th centuries CE, the Buddhist Period and Ngari-Khorsum Period between 9th and 13th centuries CE, the post Ngari-Khorsum Period between 13th and 15th centuries CE, and the Namgyal Period between 16th and 19th centuries CE. Rock art studies indicate earlier sporadic, perhaps temporary, or occasional, nomadic style settlement within Ladakh from as long ago as 700,000 years before present, involving pre-*Homo sapiens* humanoid species such as *Homo heidelbergensis*, *Denisovian*, *Homo erectus*, and Neanderthal peoples. It wasn't until a much later *Homo sapiens* period, at around 11,000 years before present, that permanent settlements arose within Ladakh, involving agriculture, transhumance, and hunting lifestyles. Trade and communication with wide geographic areas during the Iron and Bronze ages led to cultural, artistic, and economic interchanges, with Central Asian people from Mongolia and Tibet, the western Trans-Himalaya, and more westerly Central Asia. Readers are directed to Deshpande (2024) and references therein for further reading.

FIGURE 8.2 A mani wall close to Matho village. *Source*: Photograph by Lanka Nanayakkara. Note: The left-hand side photograph presents a view of the wall length. The right-hand side photograph is a close-up of a mani stone, with the inscription of a carved sacred mantra, in Tibetan script.

HISTORICAL PERIOD WITHIN LADAKH

Historically, over the past two millennia or so, Ladakh has experienced settlements and/or incursions predominantly by people described as *Dards*, originating from northern Kashmir, Gilgit, and Chitral, *Baltis* from Baltistan, and *Monpa* people originating from Tibet and Central Asia, along with other ethnic groups (see below for further details). From around the 1st century CE, Ladakh was part of the Kushan Empire, which stretched westwards into Gandhara, Bactria, and Afghanistan, and northwards to Xinjiang. This era brought Buddhism from Kashmir, although many/most followed older, more traditional Bon and other indigenous religions. Tibet extended its empire into Ladakh and Gilgit from the 7th–8th centuries CE. In 842, a Royal figure from Tibet, named Nyima-Gon, founded the first Ladakhi kingdom, recorded as the Ngari-Khorsum Period (9th–13th centuries CE), with its centre located east of modern-day Ladakh. The Nyima-Gon dynasty strengthened Buddhism in the area, but looked to Kashmir, rather than Tibet, as its key source. As Kashmir became Islamic, from the 13th century CE, Ladakh looked to Tibet for its primary Buddhist support. Increasing incursions into Ladakh from the 13th century brought instabilities and variable influences between Muslim forces to the west, and Tibet to the north and east. Ladakh became divided, with King Takpabum (from Basgo) ruling Lower Ladakh, and King Takbumde (from Leh and Shey) ruling Upper Ladakh. King Namgyal (1555–1595 CE) developed a dynasty that survived for a number of centuries, reuniting and extending the kingdom of Ladakh. Invasions from Baltistan Muslims, under the strong leadership of Ali Sher Khan, led to considerable Balti influence and the destruction of numerous temples within Ladakh. Sengge Namgyal (1616–1642) restored much of Ladakh to its previous powerful times, and built monasteries (Figure 8.3), and palaces at Shey and Leh. A statue of King Sengge Namgyal, known as the *Lion of Ladakh*, proudly sits at the gateway of the city of Leh. The Mughal Dynasty in India began to have increased influence in Ladakh with occasional Tibetan

FIGURE 8.3 Like many Gompas, Stakna Monastery sits within a stunning landscape, standing above the River Indus and commanding a strategic 360° view. *Source*: Photograph by Lanka Nanayakkara. Note: Stakna Gompa is a prime example of Ladakhi-Tibetan architecture, built during the stronger dynasties of Ladakhi rulers.

incursions. From the early 19th century CE, the Mughal Empire had collapsed, and the Sikh and Dogra rulers were vying for power in the Jammu-Kashmir region. Ladakh came under Dogra rule from 1834. King Tshespal Namgyal was deposed and exiled to a palace in Stok, which can be visited to this present day. Ladakh became part of the Jammu and Kashmir state, headed by its Maharajas, within the British Empire. Ladakh was ruled as a *Wazarat* whose headquarters were based in Skardu, Kargil, and Leh.

Following the exit of the British from India in 1947, Ladakh became a precarious, strategic zone, situated between the modern states of India, China, and Pakistan. Regular wars/military disputes between Pakistan and India and India and China arose over disputed territory between Ladakh, Kashmir, and Baltistan, and Ladakh and Tibet. This situation led to the arrival of a large contingent of Indian military forces and the construction of numerous military bases within Ladakh. Areas of former Ladakh are now situated within China, and a small area of former Baltistan, around Turtuk, is now situated within Nubra-Ladakh. Trade routes that had existed between Ladakh, Baltistan, Hunza, Gilgit, and Chitral were closed, alongside routes that straddled Tibet and other parts of Central Asia. Tibetan refugees arrived and were settled within Ladakh, following the Chinese invasion of Tibet. Ladakh was rarely comfortable during post-Independence times, with its political status as part of Jammu and Kashmir. There were strong sentiments of neglect and resentments relating to the perceived overly zealous influence of Kashmiri Muslims. Violent Buddhist-Muslim riots in 1989 and increasing advocacy of the Ladakh Buddhist Council led to the development of the Ladakh Autonomous Hill Development Council in 1995. On 31st October 2019, Ladakh became an independent Union Territory (of Ladakh) within federal India, finally separating from Jammu and Kashmir and receiving an element of autonomy it had aspired to for decades. Given the history of Ladakh, particularly over the past 1,000 years or so, it is hardly surprising that a wide range of ethnic and language groups live together within the Union Territory of Ladakh. The largest ethnic group is termed Bot/Boto, of Tibetan origin, while Baltis (from Baltistan) are particularly numerous close to the Pakistan border. Dard, Brokpa or Drokpa peoples of Aryan descent live within parts of the Indus valley. Changpa people live within the highest areas of Ladakh: in Changthang and Pangong. Purigpa people are most numerous within the Suru Valley/Kargil area. Mon and Garra people work in different parts as musicians, blacksmiths, and artisans. Tibetan refugees now comprise a group of their own identity within Ladakh. Readers are directed to District, Leh (2026), and Kaul, 1992, 1998, for further details. Ladakh opened up to non-Ladakhis, tourists, and visitors from 1974. Lower numbers of intrepid visitors arrived during the 1970s, 1980s, and 1990s. The 21st century has witnessed an exponential increase in visitors, now numbering twice as many as resident people every year. A formerly isolated and cut-off region now experiences unprecedented growth in road building within Ladakh and between Ladakh and other parts of India. There are concerns relating to over-tourism and a perceived potential of non-Ladakhis bringing large amounts of inward investment, resulting in unplanned, inappropriate, tourist-oriented development. After experiencing 700,000 years of isolation as humanoids and 11,000 years of isolation as *Homo sapiens*, Ladakh is now experiencing a time of previously unknown levels of connectivity and social change.

MOUNTAIN AGRICULTURE AND WATER ENGINEERING, LADAKH-GILGIT-HUNZA STYLE

Traditional agriculture within Ladakh (Figure 8.4) exemplifies the strong influence of Earth's geological forces within the essence of ways of life. Tectonics produced a vast continental-scale plateau and mountain region within the hinterland of Asia. These physical topographic structures then experienced the forces of geological erosion and deposition in the form of ice, water, wind, and gravity, which cut deep valleys and sharpened mountain peaks.

Within this arid montane environment, water management is essential for survival and agriculture: for all sentient beings, not only humans. Water is locked up within ice reservoirs as glaciers, within

FIGURE 8.4 Upland Ladakhi farmer milking a yak in Zanskar. *Source*: Photograph by Lanka Nanayakkara.

continuous year-round river-water supply in the form of large rivers such as the Indus, Suru, Tsarap, and Zanskar, within more ephemeral smaller rivers that occupy upland valleys, and as groundwater within alluvial fans and alluvial sediments. Most settlements do not use the mighty River Indus or similar rivers, which, to many outsiders, may be a surprise. Why do Ladakhis hardly use the largest water resource within Ladakh? The answer centres around gravity. Most settlements have developed within locations situated far above the major Ladakhi rivers. To move water from a lower water source to a higher location means that humans must work against nature. Affluent societies think nothing of this, as their large populations and high-consuming lifestyles demand industrial-scale water management. Their solutions involve major engineering works, pumping systems, dam and canal construction, and the high consumption of energy in the form of oil, coal, nuclear, and green energies required to power large-scale water-engineering solutions. Pre-Industrial Western societies worked with gravity, not against it, channelling rivers to increase their power, or utilising existing natural waterfalls, and turning water wheels for flour grinding and milling, alongside garment and other industrial activities. Highly engineered solutions are not an option for sparsely populated mountain regions. Ladakhis prefer to use water sources that are higher than, or adjacent to, villages and agricultural fields. These water supplies are in the form of glaciers, small valley rivers and streams, and groundwater via well construction. Water channels are hand-excavated and maintained through communal effort over the years and decades. Channels can transport water over many kilometres, hugging mountain sides, cleverly created to fall gently in gradient, allowing water to move under the force of gravity, from an up-valley source to villages and fields. Authors have witnessed some of the largest hand-constructed mountain water channel systems in Gilgit and Hunza, as well as in Ladakh. Engineers from industrial countries visit and marvel at the technical ingenuity of water transport systems that function without pumps, or oil, or electricity. How on earth were these constructed so carefully and smartly the visitors asked? One author had long conversations with an academic water engineer visiting Gilgit from a British University, back in the 1980s. Every evening, after returning from field visits, the engineer exclaimed remarks such as:

> I cannot believe what I see and witness. Every day I see numerous sights that just plain astound me. I see water channels carved out the side of mountains, 1–2 metres deep and 1m wide, with channel walls

> constructed from natural rock (scree, boulders, gravel) and cemented with minimal amounts of mortar. I am accompanied by five or ten or twenty local men. They tell me how whole villages or a number of villages work together to create and maintain water networks. They communally design routes with just the right gradients. I see large volumes of water transported at surprising velocities, not too fast to hurt channels, not too sluggish to impede water movement, or allow mud to settle and fill-in channels. Time is regularly put aside to maintain the channels. Rotas of labour are decided amongst the community. Once water arrives at villages and fields, a series of earthen-dykes move the water around the village area for domestic usage, and irrigation. Water can be dammed/stopped, and moved across the fields, through intricate earthen structures, supplemented by recycled old clothes and similar material, to minimize leakages. This group of people, through the Aga Khan Foundation, asked me to come and help them. We looked at their operations together from all angles. They shared their frustrations and challenges. I was struck by the fact that these people insightfully understand the large and the small scale of issues they face, but also had a number of solutions they collectively discussed and weighed-up. Frankly I felt helpless. What on Earth could I offer these sustainable water engineers of global calibre? The only solutions I can offer that may assist in small ways involve pumps, and energy, and engineering that requires maintenance from outside skills. This wouldn't help. It would undermine a system proven for centuries. I wish I could take this technology back to my world. I am blown-away by the strengths I see, that offer solutions for us as a way to live, and are so-friendly to Earth and its resources. They pool labour. They work together. They discuss and decide communally. They are driven by social-good and minimize wastage. People are respectful of every aspect of nature, from worms and small insects, to the birds and wild animals, to plants and flora, and even the rocks and minerals of the substrate and mountains. No money passes hands for labour: every minute of labour is freely given. And together, they all benefit in sharing water, which they fully understand as in short supply, and must be carefully managed.

Ladakhi and neighbouring mountain communities are world experts in water supply, water resources, and all aspects that impinge on supply and demand within their environments. The limited water resource within arid environments is a strong constraint on population sizes and the amount of agriculture that can successfully and sustainably be realised. The population of Ladakh was estimated at 118,000 in 1901, 150,000 in 1941, 275,000 in 2013, and is estimated to rise to 316,000, according to data from various censuses, including the most recent official one (Office of the Registrar General & Census Commissioner, India, 2013). Populations further back in time are difficult to evaluate. but were possibly between 50,000 and 100,000, pre-20th century CE. looking back 1,000 years. Whilst people lived within traditional villages and agricultural lifestyles, water resources have been proven to be sufficient, except during periods of exceptional drought. Lifestyle methods such as storage of agricultural products for long winters, and polyandry, limited and served population numbers. Villages developed only where water supplies were sustainably available. Many drier valleys exist and were not permanently populated, although resources within valleys may have been utilised seasonally. More recently, some villages have been abandoned as water supplies have become increasingly unreliable, perhaps due to climate change impacts. Retreating glaciers reduce water supplies locally. Ice stupas (as described in chapters above), or artificially produced ice mounds/pyramids produced during winter months, are used during the warmer season to supplement water resources. Mass tourism since 2000, especially since 2010, has significantly increased water demand. Future large-scale water supply solutions may turn to larger-scale industrial-style engineering, utilising River Indus or other larger river resources, pumping water uphill, against natural gravitational forces, and the strong tide of history.

Ladakhi soils, as with all soils, are products of rock-weathering and organic processes. Geology, topography, climate, vegetation, and local ecosystems all interplay to produce soils. Deep, high-fertility soils, regularly replenished with nutrients, are found mostly within extensive lowlands. Soils of the Indo Gangetic Plains, regularly flooded by Himalayan-derived silts and muds, feed billions of people. Mountain soils tend to be thin, less organic, rich in lithic clasts, and of low productive value. Ladakhi soils are mostly shallow to very shallow and vary widely in terms of colour, texture, and soil structure. Many soils, particularly close to mountains, are 'immature' with large contributions from inorganic local scree and gravel deposits. Cold climates only produce soils slowly, as physical weathering, rather than chemical weathering, dominates. This situation contrasts with tropical climates, where vigorous chemical weathering produces deep soil profiles. Soils that form on limestones, for example, within parts of Zanskar, have higher, more alkaline pH characteristics with free calcium carbonate. Soils that form on igneous and metamorphic rocks are also mainly alkaline due to highly evaporative climatic conditions during summer months. Most soils have pH levels of 7–9. Some researchers, such as Gupta and Sanjay (2017), have classified soils as: (1) those that occur on slopes or proximal to slopes and are somewhat inorganic in character. These soils are termed *entisols*; (2) thicker versions of entisols, termed *inceptisols,* develop deeper soil horizons; (3) thicker, darker soils with organic-rich surfaces found within valleys are the most fertile Ladakhi soils, termed *mollisols*; and (4) *alfisols,* which become leached (depleted in nutrients) and somewhat clay-rich, are associated with woodlands and forests within mountains. Soils are also classified as sandy or sandy-loam (containing more humus and silt), rich in potassium but depleted in zinc, iron, phosphorus, and manganese. Regularly cultivated soils can be organic-rich due to the application of manure and fertilisers.

Land in Ladakh is cultivated at heights of between 2,500 and 4,500 m asl (Figure 8.5). Only around 0.2–0.3% of the total land area of Ladakh is under cultivation, equivalent to 20,000 hectares. Food grains (barley and wheat) and pulses comprise over 70% of all crops, with the remainder devoted to fodder crops and vegetables. Fruits and nuts such as apricots, sea buckthorn, apples, walnuts, peaches, grapes, and pears are grown, particularly at lower altitudes around the Indus Valley, Leh, Kargil, and parts of Nubra. Ladakhi livestock farming involves the rearing of yak, Ladakhi cattle, dzo (a cross-bred animal between yak and cattle), pashmina goats, horses, donkeys, and camels. Most milk is turned into butter. Yak and Ladakhi cattle milk production is relatively low. Milk has a rich texture. Livestock graze around villages, and are regularly taken to pastures on higher ground. Wolves, bears, and snow leopards pose risks to livestock. Livestock are kept close to, or even within, village houses during the long, cold winters. Storage of fodder is considered as important as human food, as meat is essential for times of year when food can become scarce. See Singh (2025) for further details concerning Ladakhi agriculture.

Helena Norberg-Hodge wrote an international best-selling book and influencer called *Ancient Futures* in 1991, centred on traditional and changing Ladakh, as well as founding a non-government organisation, based in Leh, called *Ladakh Ecological Development Group*, a part of Norberg-Hodge's international *Local Futures* initiative. Norberg-Hodge first came to Ladakh in 1975, shortly after Ladakh was opened to outsiders. Ladakh was only connected to post-independence India by road in 1962, from Srinagar, Kashmir. Norberg-Hodge's accounts in *Ancient Futures* capture the culture of

FIGURE 8.5 Whilst barley fields are filled with crops during the short growing seasons, they lie fallow for much of the year, often iced over during the coldest months. *Source*: Photograph by Lanka Nanayakkara.

Ladakh relatively unaffected by the outside world. She stayed in Ladakh for two continuous years and spent half a year for a further 17 years in Ladakh. She witnessed change and lamented many of the impacts of change, while recognising that things can't stay the same forever. Norberg-Hodge is a critic of what she sees as the more individualistic, competitive, stressful, lonely, environmentally unfriendly approach of what she terms the *industrial society*. While not meaning to overly romanticize traditional societies, understanding they have their challenges and aspirations, Norberg-Hodge would like to see a wider world develop more along the lines of inter-dependency, with increased collaborative and communal spirit, and deep respect for all things natural, characterised (as she viewed) within traditional agricultural Ladakhi society. Accounts below of Ladakhi agricultural culture and society are drawn partly from Norberg-Hodge's *Ancient Futures*, alongside the authors' own experiences within Ladakh over many years.

Many Ladakhi villages comprise 50–200 thatched stone, mud, and cement houses, two to three storeys high, that straddle roads and streams, set within a patchwork of fields. Most villages are situated within intra-montane valleys, some quite isolated and high up, located a number of kilometres distant from the main road, whilst other villages are located close to, or within, larger valleys such as the Indus. All villages have gompas (or temples), and many have communal meeting house facilities. Surprisingly, branches of banks exist within even tiny villages. Living within a village house is communal, with much time spent in the kitchen area, close to the fire, keeping warm during the long, cold winter seasons. Animals are an integral part of home living, housed close by during winter seasons. Yaks, dzo, horses, cattle, and sheep may be housed in pens with high walls: broken glass or similar embedded within the tops of walls, aimed to deter bear attacks and similar. Meals are communal. Rooms are dark, with small windows, preserving heat in winter. Floors are covered with numerous woollen and sheepskin mats. People sit on cushions next to low tables, called *choktses,* on which food and drink are served. Ladakhi (or Tibetan) salted butter tea, called *namkin chai or gur-gur chai*, is a brew made from green tea, salt, and yak butter, blended within a long wooden and brass trumpet-shaped churner, called a *dombo* or *chamdong* (Figure 8.6). Ladakhi butter tea is a high-calorie beverage that delivers warmth, hydration, and salt replacement. It could be viewed as much like a soup as a tea, and is an acquired taste for outsiders. Food revolves around barley and wheat. Roasted barley flour called *tsampa* is the ubiquitous central ingredient of traditional Ladakhi food (Figure 8.7). It can be consumed as it is, or tsampa can be mixed with a liquid such as tea or hot water, to create a paste, or bread-like dough, which is widely eaten. Barley is also processed to produce *chang* or barley beer, widely drunk at social occasions, with meals, or during long days in the fields. Other staple foods include *momos*, steamed dumplings made from wheat or barley flour, stuffed with vegetables or meat, or both, *trukpa*, a noodle soup or broth made with cabbage, potatoes, vegetables, and/or meat, barley or wheat flatbreads, local cheese, fruits and nuts, and sweet dishes. Most Ladakhis carry bags of tsampa on long walks and mountain treks, or travelling within the larger valley lowlands, as it provides ready-made nourishment on the go, and is always there to share.

Ladakhi agriculture follows annual cycles governed by seasons and climate. The cycle begins between February and June, dependent on altitude. Sowing, irrigating, and harvesting were traditionally communal activities. Draft animals were widely used, including donkeys, horses, and yaks/dzos. Appreciation of all aspects of the environment entails a range of rituals and prayers, paying homage to local deities of the land, the water, and the air. Monks may be involved in more intricate ceremonies at different parts of the agricultural cycle, ensuring that the local environmental elements, spiritual and material, are in balance, praying for a good harvest for all, with minimal adverse impacts on insects and animals. Manure is brought to the fields by donkeys or other draft animals and spread by villagers. Fields are ploughed with a wooden plough and dzo animals. Once seeds are planted, watering is the main activity, undertaken largely through irrigation dyke systems, manually operated throughout the year, to ensure water gets where needed across the field network. As water is so precious, leakages are avoided at all costs.

Cutting of crops was undertaken manually, and still is in parts. Villagers turn out en masse to ensure that all crops are harvested optimally. Tasks are not viewed as burdensome. Work in fields is essential for the community, with the forthcoming cold winter season in mind. Chang is consumed during work in

FIGURE 8.6 Butter tea or *namkhin chai* is mixed within a traditional churner called a *dombo or chamdong*. *Source*: Photograph by Lanka Nanayakkara.

FIGURE 8.7 Monks make ritualistic sculptures from tsampa and used in Pujas. *Source*: Photograph by Lanka Nanayakkara.

the fields, and people sing traditional songs as they engage in their agricultural activities. Norberg-Hodge quotes song lyrics she had heard whilst she accompanied villagers in the fields. Two examples give an impression of song lyrical content with respect to agriculture:

'Manjushri, the embodiment of wisdom hark!/ The gods, the lhu, and owner spirits of Mother Earth hark!/ May a hundred plants grow from one seed/ May a thousand plants grow from two seeds/ May all grains be twins!/ Please give enough that we may worship the Buddhas and Bodhisattvas / That we can support the Sangha and give to the poor',

and;

'I shall not drink more chang / Only if someone can take the blue skies on his lap will I take chang/ The sun and the moon take blue skies in their lap/ Drink cool chang. Drink! / I shall not drink more chang/ Only if someone can braid the water of streams will I take chang/ The fish with golden eyes braid the water of the streams / Drink cool chang. Drink!'.

These lyrics express elements of agricultural and village life, including respect for the interdependence between people and environment, the desire to harvest abundant crops not only for the villagers or the farmers, but with excess to share with the poor and the Sangha (monks), the communal joy of working together and sharing chang (beer) together, and the playfulness and joy in camaraderie. Once crops were harvested, they are threshed within a circular hardened mud ground with a stick at the centre. Cows and dzos are tied to the stick and walk around in circles, treading and threshing the harvest. People walk behind the animals with baskets to catch any droppings that may fall from the animals. People fling grain into the air, separating the grain from the chaff. Crops are carefully stored for humans and animals alike. Many villeages take the more commercial crops grown to market, in places such as Leh (Figure 8.8). All agricultural and village practices respect resources and minimise waste. In many ways, there is no such thing as waste in traditional lifestyles. Barley used to brew chang is used for fodder after brewing. Clothes that have been patched so much they can no longer be worn are used within irrigation channels to prevent leakages. Human toilet waste from dry latrine-style toilets is collected annually, to be added with animal manure for fertiliser.

Groups of villagers herd their animals to higher pastures, as required (Figure 8.9). They will pack supplies in bags and strap these to ponies and donkeys. They take flocks/herds of sheep, goats, and yaks to higher pastures and may trek for a whole day or several days using mountain tracks and high passes. They have built rudimentary houses or shelters and pens within the richer, more extensive pastures and spend several nights at a time here. At night, animals are brought into pens for protection against predator animals. Animals may spend time in high pastures for several months. People look after them on a rota basis. Increasingly, nomadic ways of life, such as in Changthang, involve transhumance activities for whole summer periods.

Agricultural practices vary in modern-day Ladakh. Traditional agriculture remains in many places. The advent of tractors and threshing machines has reduced the need for draft animals in places. More individualistic, rather than communal farming, is more apparent, although cooperative culture remains strong. Many people have left their villages for Leh and set up businesses or taken up employment in town. Leh has dramatically expanded. Ribbon developments extend along roads in all directions from Leh. Leh's population in 1961 was around 44,000, 117,000 in 2001, and 133,000 in 2011 (Office of the Registrar General & Census Commissioner, India, 2013). Particularly rapid growth of tourism from 2011 to 2015, continuing needs to service military personnel, and increased administrative and urban employment requirements have all fuelled urban drift, depopulating numerous rural areas. Many in Leh maintain property in their villages and rent out their farmland to tenant farmers. The lower need for farm animals has led to a significant increase in stray animals, as families and villagers can no longer look after them. In Taru, west of Leh, we spent time whilst undertaking our geological work in 2019, with a farmer called Tsering Dorjay. We had heard about his charitable activities, caring for stray animals. He told us his story, part of which is written up by Chumikchan (2019). We came across a large barn with many animals inside and others grazing across the valley, although pickings were limited as it was a winter season, with snow on the ground. Cows and dzos predominated, but we also saw horses, donkeys, sheep, and goats. People in the valley had told us about Tsering with great admiration, and we had the opportunity to meet up with

FIGURE 8.8 Agricultural produce brought to market in Leh. *Source*: Photograph by Lanka Nanayakkara. Note: The market seller is dressed in traditional Ladakhi clothes, turning a prayer wheel as he works.

FIGURE 8.9 High mountain pastures for Ladakhi yak herds. *Source*: Photograph by Lanka Nanayakkara.

Tsering himself. Tsering showed us around his facilities, his barns, and other outbuildings, together with the range of animals he cared for. All were in sound health and spirit. We were invited to Tsering's house, met his family, and were given butter tea and incredibly tasty home-made Ladakhi bread and butter. Tsering communicated his history in an animated fashion:

> Thank you, both of you, for showing interest in my project, and allowing me to narrate my story to you. For many years I was a stone mason and worked far and wide across Ladakh. I witnessed changing times. Back in time people lived more traditionally within villages high-up in mountain valleys. Lives revolved around farming and the cycle of agriculture. Animals were an integral, core, essential part of this way of living. Animals gave milk, hides, cloth, meat, and their labour in the fields. People were so proud of their animals; they knew every animal personally. Yaks, cows, dzos, and sheep are killed for meat and food when required, especially during cold winters and vegetables are scarce. But this is not taken lightly. Older animals tend to be killed, as they have at least lived their lives. Before any animal-killing takes place, prayers are recited, animals are calmed, great respect is given, and the sacrifice of the animal for the welfare of humans is clearly recognized. We ask that these animals gain many merits, amplified a million times by giving up their life for us, and through these merits gaining higher rebirths. Ladakhis and Tibetans prefer to kill larger animals if they must kill, as a larger animal provides much food for one life. It is forbidden to catch fish or any aquatic animals in Ladakh, as we believe that it is less wise to take lives of many smaller creatures for our welfare. Domestic animals are treated kindly with utmost responsibility. They have good lives. Meat is not consumed on an industrial scale, and large quantities are not eaten for most meals. Many meals in Ladakh are vegetarian. However, I saw how things changed. Agricultural lifestyles became less important for some, seeking livelihoods in Leh, in India, and beyond. My own valley, Taru, close to Leh and now impinged upon by urban sprawl, is a clear example of change. Some houses now lie empty. Some fields now lie uncultivated. Other farms are rented out to tenant farmers. Agriculture is less communal and more mechanized. Wherever I went in Ladakh I saw animals drifting around, with their ribs showing, in poor health, uncared for. I then saw trucks filled with live animals heading to Kargil for mass slaughter. And I finally witnessed animals from my own village, Taru, heading for the same fate. My heart broke. I couldn't live with this situation without doing something. I asked and was kindly granted permission from my neighbours to use some local land for stray animals. I built what I could afford in terms of accommodation. People started to deliver animals: similar stories to my experience were relayed to me. People struggled to look after their animals. Looking after animals is a year-long, every-day activity, with a high level of commitment. During past times this was not a burden, nor thought of as such at all. On the contrary, it was an honour to care for our animal friends. Modern life brings different demands, within a work location faraway from fields and animals. Not so long ago Leh was small, and within a twenty-minute walk in any direction you would be in fields. Even Leh people were connected directly to the land. Today Leh has grown, and urbanised people have become detached from agricultural lifestyles, with no time for animals. Young people can be less interested in taking up the agricultural village life of their ancestors, instead seeking brighter lives in Leh and beyond. So, animals arrived from far and wide. After a while they became so numerous it was overwhelming. I now ask that for every animal given, there is also a commitment to pay for their food and upkeep. Even so, my heart breaks with every stray uncared animal, so I take on more. My animals freely roam and graze during the day and come to the barn at night. I found it difficult to work and care for the animals and keep up my day job. I gave up my work. I now run a charity and rely on donations. It is of course very expensive to care for so many animals. So far people have kindly donated what I need. But it's always a challenge to attract funding. In Buddhism animals are considered as sentient beings, just as important as humans. Like humans they are part of Samsara, the realm of cyclic rebirth. I was once an animal, now I am a human...I may be an animal in a future rebirth, although of course I hope for different destinations. We must show compassion for all animals of all sizes and shapes. Of course, we must kill some to live. Even vegetarian lifestyles involve the death of many insects and beings. I consciously further develop a deep respect for animals. and minimize the sufferings my lifestyle entails to animals. Caring for these stray animals is such a joy. They are so healthy and happy and every moment gives me gifts I will never forget.

We left Tsering in pure awe and admiration: to witness such dedication, commitment, and compassion: so humbling and inspiring.

LADAKHI TRADITIONAL MEDICINE

As with many indigenous or traditional societies, Ladakh has a thriving traditional, ayurvedic, here termed *Amchi or Sowa Rigpa* medical culture. Being holistic in nature, this style of medicine does not treat 'patients' and their ailments as separate individual issues, but rather as part of the connections a patient has to their moods and life history, their family and community situation, and to the environment around them. Sowa Rigpa philosophy treats the holistic well-being of a person, with the aim of improving all aspects of life, health, well-being, and longevity. Patients are examined in terms of the balance of internal energies termed wind or *loong,* bile or *tripa, peken* (phlegm), heat or *mkhris-pa*, and cold or *Bad-kan.* The character of the pulse of a patient is perhaps the most critical part of a traditional doctor's or *Amchi's* methodology. S/he carefully senses the pulse of the patient in front of them with their own hands, astutely observes the demeanour and disposition of the patient, asks numerous questions, takes her/his time to sense the situation, and decides on a course of action, providing a diagnosis and prognosis. If energies are in balance and healthy, a healthy situation is diagnosed. If an energy imbalance is detected, the situation is thoroughly explained, and links between the health condition the patient is suffering, their lifestyle, their relationship issues, and links to environmental factors such as heat or cold or dust exposure are analysed. The patient is usually prescribed traditional herbal medicine and provided advice regarding fluid and food intake, times of eating and drinking, various activities, and observing her/his bodily system through toilet or phlegm products.

The *Amchi or Sowa Rigpa* traditional medical system is formalised and accredited to a high level of official recognition. Historical roots for *Sowa Rigpa* may extend back 3,000 or 4,000 years, incorporating Persian/Iranian, Greek, Mongolian, Indian, Chinese, and Tibetan knowledge alongside Bon and Tibetan Buddhist understandings. The National Institute of Sowa Rigpa (https://www.sowarigpainstitute.in/) provides education and medical services for all, engaging in training, education, certification, and a wide range of research and development activities underpinning traditional Ladakhi medicine. Graduate degrees up to PhD are offered. The Men-Tsee-Khang Sowa-Rigpa Medical College & Hospital, Dharamsala (https://dcollege.mentseekhang.org/college/) was developed by the Dalai Lama in 1961 and comprises a university, hospitals, and community services to the local and wider Tibetan and Indian communities throughout India. This facility, with roots back to Tibet and thousands of years of medical knowledge, actively supports further development of Sowa Rigpa medicine across India and beyond. Historically, Ladakh was closely connected with Tibet, maintaining live connections with cross-fertilisation in numerous areas of knowledge, including agriculture, trade, medicine, and religion/philosophy. Since the invasion of Tibet in 1959, these close cultural links were severed. The creation of Tibetan-Ladakhi medical institutes within Ladakh and other parts of India are examples of a new reality, with Tibetan cultures finding ways to become self-sufficient in exile. Readers are directed to Chahuan et al., 2020, and Ladakhdekho.com, 2026, alongside web links to the Ladakhi and Dharamsala medical institutes for further reading.

The *Amchi* medical system has a high level of respect within Ladakh. Most people regularly consult *Amchis.* Historically, *Amchis* are part of a familial tradition, with knowledge passed on between generations over thousands of years. *Amchis* are embedded within communities, serve communities, and are a respected part of communities. They build their reputations within their communities. Those with high reputations are highly sought after. The Sowa Rigpa medical system works in parallel with 'Western' medicine. Government hospitals and clinics are present in many parts of Ladakh. A hospital or clinic may offer Western and traditional services in parallel. Remote villages rely on skills within their own communities, as seeking help elsewhere may involve long hours or days of travel. Medicinal herbs and medicinal plants present in Ladakh are closely linked to mountain ecosystems, which, in turn, are linked to the tectonics that created the mountains: all things Ladakhi are ultimately rooted in the tectonics that built their landscapes.

LADAKHI DRESS, MUSIC, AND DANCE

Ladakh hosts numerous festivities of various sorts. Some revolve around village activities, linked to marriage or the agricultural calendar, and important family and village events. Some are religious, linked to the many monasteries/gompas of Ladakh. Some are linked to public holidays and national/regional days. Many festivities are hosted by monasteries at different times of the year, often involving complex ritualistic dancing by the monks, dressed in masks and traditional costumes. Dances and music relate tales from Tibetan Buddhism and the fight between good and evil. Whilst it's difficult to capture the full variety of events, this section gives a flavour of monthly/annual festivities in Ladakh. For further details, readers are directed to Bedi, 1986, Neetu, 1994 and Shakspo, 2024.

Ladakh holds annual festivals in Zanskar, Leh, and Norbu, showcasing music and dance from all areas of Ladakh, including those from Muslim areas such as Kargil and Suru Valley. These events involve hundreds of dancers and musicians, attracting thousands of people from all over Ladakh, India, and internationally. All performances involve traditional dress with different styles from different regions. Musicians back dancing performances with a Ladakhi orchestra and live music. Musicians are often from the Mon people and practise wandering lifestyles: such people are termed *Bedas.* Royal families and nobles of old employed residential musicians. Other musicians wander from village to village, festival to festival, and monastery to monastery. A Ladakhi orchestra involves instruments called *daman* and *surna* (kettledrums and flageolet), stringed instruments termed *damnayan,* and flutes called *limbu.* At the national festivals, the musicians arrange their instruments around a raised stage. Dances involve groups of people, both women and men, or women/men only. Each dance is backed by rhythmic and melodic music. All dances originate from specific villages or areas, and performers practise for a long time before festival performances. In villages, dances and songs are part of everyday life. Dances have themes that vary from sowing or harvesting crops to the behaviours of birds and wildlife, to love, to religious stories and themes, and everyday life observations. Movements are slow, involving careful arching and swaying of hands and upper body, and reaching down to the stage. Slow, mesmeric movements and accompanying music are rather meditative to watch.

Traditional dress is on show at these festivals (Figures 8.10, 8.11 and 8.12). The same dress can be seen at numerous village events. Ladakhis are famous for their clothing styles, attracting people from across the world to see authentic dress quite different from the 'normal-western' fashions across the world. It is refreshing to see many Ladakhis still maintaining their dress customs, although a move to ubiquitous 'western style' dress is also prescient. Traditional male fashions include an outer coat or gown called *Gonchas* or *Kos,* made from wool, coloured burgundy red or grey with yellow or gold trimmings. More colourful, embroidered silk Gonchos are worn by some. Women wear more flowing outer robes called *Sulmas,* made from wool, brocade, silk, or velvet, and can be brightly coloured. Sashes called *skeyraks* wrap around the waist of the Sulmas. Ladakhi hats or *tipi* are particularly characteristic. It's a fabric pill-box hat, with the lower part out-turned, spreading radially from the central pill box. Moccasin style, colourful shoes with upturned shoe fronts called *pabu* are made from wool or yak hair. Warm capes called *yogar* and *lokpa* are worn as outermost garments. Jewellery is made from silver, gold, coral, lapis lazuli, and turquoise, mostly in the form of earrings, necklaces, amulets, and bracelets. The most characteristic Ladakhi jewellery piece is a highly ornamented head-dress called a *perak.* Peraks (Figure 8.13) are highly prized bridal head dresses but worn on many occasions, and are densely-studded with gem stones, as noted above, but could also include garnet and shells.

FIGURE 8.10 Traditional Ladakhi female dress. *Source*: Photograph by Lanka Nanayakkara. Note: This lady wears an outer burgundy coloured dress called a *sulma*, and a traditional pill-box hat called a *tipi,* adorned with jewellery made from silver, gold, turquoise, lapis lazuli, and coral.

FIGURE 8.11 Pilgrims dressed in traditional Ladakhi clothes, with prayer wheels in hand, are walking towards old white stupas or chortens adorned with prayer flags. *Source*: Photograph by Lanka Nanayakkara.

FIGURE 8.12 Ladakhi man dressed in traditional Ladakhi dress, Leh market. *Source*: Photograph by Lanka Nanayakkara.

FIGURE 8.13 Beautifully adorned Ladakhi head dresses called *peraks,* which are culturally highly valuable, studded with lapis lazuli, turquoise, and other precious stones.

BUDDHISM IN LADAKH

Whilst acknowledging the presence of other religions within Ladakh, particularly Islam, but also Moravian Christianity, this section will focus on Ladakhi (Tibetan) Buddhism. For the great majority of visitors to Ladakh, alongside the mountains and physical beauty, it is the Tibetan Buddhist culture that is by far the biggest attraction. It is Tibetan Buddhism that has given Ladakh a unique culture beyond Tibet proper and marked the landscapes with the many signatures of Tibetan Buddhism (Figure 8.14).

The Himalayas and Tibet are renowned, famous, and globally admired as places of spiritual endeavour and advancement. Mystical, meditative traditions are the ones that attract most attention and curiosity. People come from around the world to be a part of, to witness, or merely to observe, the deeply philosophical traditions of the Himalaya. Why did this happen, here, in this place? What is it that shapes the human mind to seek for the deepest answers to the most mysterious questions? Spiritual traditions and Indigenous spiritual beliefs are strong in many mountains of the world, from the Bolivian Andes to the North American Rockies, and the Shamans of Papua New Guinea. There is a visceral connection between mountains and spiritual explorations. We claim that tectonics has even influenced these deepest of human searches. All mountains are fundamentally the product of active, dynamic Earth tectonics and the interaction of its numerous plates. Whether it be one continent colliding with another, as in the Himalaya-Alpine chain, or the rise of large island arcs, or submarine mountains arising from the splitting of plates and massive magmatism. It is this manifestation of our planetary restless energy that provides the home for the sages, seekers, and truth searchers. The Himalaya and Tibet provide sanctuaries and reclusive, isolated terrains for solitary Sadhus. Modern times excepted, few parts could easily be reached and required strenuous walks and treks in arduous conditions for thousands of kilometres. The very nature of the sky, the atmosphere, and light changes as humans rise to half the air pressure of sea level. Existing within such hostile abodes, with limited food and water options, intense cold, and sweltering heat, is not for the faint-hearted. Many historically came to the Himalaya, made their own spiritual discoveries, with

FIGURE 8.14 Phyang Gompa or Monastery, built on top of a large terminal glacial moraine, and surrounded by outstanding Ladakhi mountains. *Source*: Photograph by Lanka Nanayakkara.

some becoming influential teachers, writers, and philosophers, forming movements that millions followed. How many religions existed throughout human history? We only know of those that exist today or are extinct, leaving traces in writings or rocks. Many more must have come and gone in long-distant human times. The historical Buddha, Buddha Shakyamuni, the Awakened One, or the Tathagata, lived as a child and young man in Nepal, not far from the Himalaya. He knew of the Himalaya and the spiritual seekers, but he himself, as The Awakened One, did not spend time in the Himalayas. The time of the Buddha in North India, some 2,400–2,600 years ago (depending on scholarly interpretations), was one of a revolution in philosophy and city development: perhaps the two went hand in hand. Vedantic, Jain, Nihilists, Materialists, and numerous other movements flourished within a period of spiritual exploration. The Buddha benefitted from this highly inspiring time; he learned from his contemporaries and the ancient Vedic traditions, but went on to develop a new path. Communities that followed the way of the Buddha arose in North India to begin with, including parts of the Lower Himalaya, and then proceeded to much of the sub-continent, Sri Lanka, SE Asia, and Indonesia. Emperor Asoka (304–323 BCE) was strongly influenced by Buddhism and became a key catalyst in the spread of Buddhism. A northern Mahayana tradition of Buddhism developed around 2,000 years ago and spread through much of India, taking strong roots within the Himalaya, Tibet, China, Mongolia, Siberia, Korea, Japan, and Vietnam. From around the 3rd and 2nd centuries BCE, Himalayan regions such as Bactria, Gandhara, Swat, and Kashmir became centres of Mahayana Buddhism. These co-existed with traditional religions such as Bon-po and other indigenous religions of the Mons, Dards, Tibetans, and Balti peoples. This sets the scene for our understanding of the influence of Buddhism in Tibet, which is picked up below (Figure 8.14). Readers are directed to Harvey, 1990 and 2004, for further details.

Buddhism has formed within Ladakh in two main waves, both of the Mahayana form. The earliest form of Buddhism probably dates back to the time of Emperor Asoka. At that time, Buddhism was well established in Swat, Pakistan, and Kashmir. Taxila, north Pakistan, was a university centre of Buddhism from as early as the 5th century BCE, and Swat, Pakistan, contains stupas from the 2nd century BCE, contemporary to some of the oldest stupa ruins in Ladakh. Large Maitreya Buddha statues of probable

Kashmiri-Bactrian-Gandharan style are well preserved at Mulbekh and in the Suru Valley, Ladakh. These statues are dated at the 7th–8th centuries CE; some argue for earlier dates at c. 2nd–3rd century CE. Buddhist inspiration and support for Ladakh came from Kashmir until around the 13th century CE, although it began to compete with Tibetan influences from earlier times. The first arrival of Buddhism in Tibet, which also influenced Ladakh, was from the 7th century CE. An immense figure in Tibetan Buddhism, Padmasambhava, arrived in Tibet via Ladakh, emanating from Swat, and spent time in Gilgit and Baltistan. At this time, Gilgit was a vibrant Buddhist centre. Ladakhi Tibetan Buddhist influences were strengthened by the arrival of Naropa, a key influencer, spending time in Zhongul and Lamayuru. From this period, Tibetan Buddhism continued to be the dominant influence within Ladakh. Today, all the main sects of Tibetan Buddhism are present in Ladakh, with the Drukpa School being the dominant sect, partly through historical ties with Bhutan at different periods of history. For further details, readers are directed to Shakspo's *A Cultural History of Ladakh* (2024) with a long section on the history of Buddhism in Ladakh.

Those less familiar with Tibetan Buddhism can find it quite exotic and rather baffling. Whilst Theravada Buddhism focuses on the historical Buddha, Tibetan Buddhism, acknowledging Shakyamuni Buddha as the source of Buddhism, also includes a pantheon of other sacred figures, including numerous Buddhas, Bodhisattvas (highly realised beings who delay Nirvana to help other beings), and deities. Wherever Buddhism has taken root, it has coexisted with other contemporary belief systems and often incorporated elements of them. In Tibet and Ladakh, the coexisting Bonpo religion (still present today) strongly influenced Buddhism, as have elements of Hinduism, giving Tibetan Buddhism a uniqueness. Whilst the core of Shakyamuni Buddha's teachings is the essence of all Tibetan Buddhist teachings, later Buddhist scholars and philosophers added tremendously to the canon of sacred texts. Tibet became particularly important by becoming the repository of many Buddhist texts, once held in India but threatened with survival by 12th and 13th-century invaders. All Sanskrit texts were also translated into Tibetan. Tibet became one of the most important centres for all Buddhist texts, holding a comprehensive archive of the Tripitaka, Pali Canon, Nalanda Mahayana texts, and Tantric Vajrayana texts. Tibetan Buddhism developed a wide variety of physical sacred elements that mark any landscape in which it exists. These include: the temples themselves, termed *gompas*, the largest of which housed (and houses) thousands of monks within university-school-temple complexes; *stupas or chortens* (a monument containing relics of the Buddha or holy beings, with a square base, a bell-shaped middle section, topped with a spire) with unique Tibetan characteristics; rock carvings of Buddhas and Bodhisattvas; prayer flags hung from many areas of the landscape; mani walls and stones (walls and stones with carved mantra inscriptions); and pyramids of whitened stones (*Lhatos*), with relics of sheep and goat skulls and similar as offerings to local deities. Many sacred works of art are created, including *thangkas* (holy paintings), statues, butter-tsampa-sculptures, mandalas (artistic representations of the Buddhist cosmos), along with ritualistic objects that include *prayer wheels, mandala beads, dorje bells,* daggers (*phurbas*), *vajras* (brass thunderbolt objects), hand drums (*damaru*), and trumpets *(dungchen* and *kangling*). Monks wear burgundy or ochre red robes, live austere, celibate lives, and spend their time undertaking a range of religious and spiritual practices, alongside numerous social-good activities. Day is spent in prayer and chanting, scripture study, performing pujas (religious ceremonies and worship), chanting sessions, and meditations. Larger universities are centres of study conferring degrees up to PhD (termed *Geshe* degrees). Whilst clergy (*Sangha*) are predominantly male, a significant number of nuns and nunneries exist in Ladakh.

Several thousand monks and around 1,200 nuns live in over 100 monasteries and around 30 nunneries in Ladakh (Figure 8.15). There are many other small temples with no resident monks or nuns. Monasteries and temples (Figure 8.15) vary from large, affluent institutions to small, one-building village affairs. The larger or more famous monasteries include Hemis, Thikse, Spituk, Phuktal, Karsha, Diskit, Alchi, Lamayuru, Likir, Matho, Phyang, Stakna, Sakti, Thaktok, Chemrey, Sankar, Rangdum, Wanla, Zongkhul, Bardon, and Lingshed. Authors have spent much time in various monasteries, attending Pujas, sharing meals, and talking with numerous monks and nuns; the account below comes from our experiences, impressions, and discussions with other Ladakhis.

FIGURE 8.15 Phuktal Gompa is set within a honeycomb network of caves within remote Ladakhi cliffs. *Source*: Photograph by Lanka Nanayakkara. Note: Note the radiating upwards outstretching array of fractures beneath the Gompa, emphasising the influence of tectonics.

FIGURE 8.16 Monks create a coloured sand mandala with painstaking effort over days and weeks. *Source*: Photograph by Lanka Nanayakkara. Note: Once completed, the monks destroy the mandala, demonstrating impermanence and non-attachment.

Monasteries and temples (or gompas) are at the heart of Ladakhi culture. Most, if not every, village has a gompa/temple or monastery. Monks do not live lives removed from communities. Rather, they are integrated within local communities. Most of the monks and nuns are from villages within close proximity to the monastery or nunnery. Village families support their ordained sons and daughters, along with other members of the Sangha. Monks attend to a wide range of village needs. Rituals and prayers related to the agricultural cycle, to the blessing of homes, to breaking ground for new buildings, to the setting up and continued work of businesses, and so forth, are performed. Traditional funerals involve serious work from the monks. They may be present with a deceased person's body for seven days before the final procession and burials/cremations. The departing spirit of a dead person is thought to reside close to the body for some time after death. Monks and lay people chant and recite prayers to help the deceased spirit find her/his way to the next life, or to leave samsara altogether. The community itself is ever-present at the monastery, helping with building work, maintenance, Sangha support, requesting pujas, and numerous other activities. Relationships between the Sangha and the community are definitely symbiotic. Whilst undertaking fieldwork, we gave numerous monks lifts to various places. They were visiting individual families for prayer and counselling sessions. A wide range of activities exists within temple complexes. Schools are present for younger monks. Kitchens involve communal effort, as does the ordering and collection of food stores. Artisans carve mani stones and statues, or paint thangkas and other works of art. For special occasions, monks stand around a large table creating a large mandala made of coloured sand. On other occasions, monks spend time creating butter and tsampa statuettes and deities for pujas. Some monks reside in hermitages for several months or years, undertaking isolated meditations. We spent time with one old nun aged almost 90 years. She lived in her own room with a magnificent view across Zanskar. She had one seat and a mat to sleep on, a single burner to cook on, and a fire-heater. It was bitterly cold in winter. The toilet was an outside latrine, and the old nun struggled with rheumatism and pain. However, she was part of a mutually helpful community of around twelve nuns, who supported one another. Large temples tend to be dark places. They are full of shrine rooms and prayer halls, thangkas, and statues, with numerous altars to various Buddhas and deities bedecked with bowls filled with water, lamps burning butter, incense perfuming the air, flowers adorning altars, and the throng of visitors. Pujas and chanting sessions are held at different times of the day. Some temples, such as Thiksey, house huge Buddha statues that people walk around at different levels. Pilgrims circumnavigate stupas or chortens in a clockwise fashion, holding prayer wheels and mandala beads, reciting mantras such as *Om Mani Padme Hum*. Many temple complexes have a pilgrim's round called a *kora*, which can extend around the perimeter of the temple grounds. Many prayer wheels are present along the kora (Figure 8.17). Pilgrims turn each and every wheel whilst reciting mantras. Larger gompas have spacious courtyards used for a range of festivities throughout the year (Figure 8.18).

One of the most remarkable and awe-inspiring sights of the gompas is their very location, as numerous figures in this chapter and throughout the book testify. Here, again, there is a direct link between mountains, tectonics and culture. Gompas are often built upon high, perched locations, overlooking valleys, villages, towns and slopes, standing proud for all to see and make an effort to walk towards. The features on which they stand all have geological significance. Some are upstanding ridges, the product of thrust tectonics, some are on top of glacial moraine hills, some atop cliffs filled with radiating fractures, others surrounded by batholith or volcanic or sedimentary rock exposures, others perched above incised river valleys. So many have extraordinarily impressive scenic locations! Each location could constitute a site of special scientific interest or a geosite. Chortens/stupas often line paths approaching gompas, and are positioned within other stunning locations around the gompa. Stupas or chortens can occur in groups or clusters, set within a range of landscapes. Near Shey, there is a 'forest' of chortens set within an open valley region, comprising tens of chortens/stupas. There is a report that historically, people who had committed crimes were asked by the local King or Noble person to construct the stupas, in so doing atoning for their actions (personal communication). Standing at many stupa sites, people often take time examining what can be a stunning 360° landscape view surrounding them. Prayer flags are another ubiquitous Ladakhi

FIGURE 8.17 Pilgrims circumnavigate *kora* or a pilgrim's path with numerous prayer wheels that stretch across the perimeter of many Gompas. *Source*: Photograph by Lanka Nanayakkara.

landscape feature (Figure 8.19). Flags vary in size from A4 page and smaller size to much larger. They come in five auspicious colours: blue, green, white, red, and yellow. Blue represents sky and space, white represents air and wind, red represents fire, green represents water, and yellow represents earth. Prayers are printed on the flags, along with images of the wind horse, dragons, Shakyamuni Buddha, Padmasambhava, Bodhisattva Tara (Dolma), snow leopard, tigers, and so forth. Prayer flags are linked together with string and placed in prominent positions: on bridges, along mountain passes, around stupas and gompas, house roofs, entrances to villages, and numerous locations within the landscape. There is rarely a view that does not include chortens and prayer flags. Flags blow with the wind, sharing their prayers, mantras, blessings and good wishes for the well-being and peace of all sentient beings. Ladakhis replace old flags with new

FIGURE 8.18 Large temple courtyard within Hemis Gompa, used for numerous festivals, dances, social, and religious events, with details of internal gompa architecture presented. *Source*: Photograph by Lanka Nanayakkara.

FIGURE 8.19 Prayer flags are an integral and ubiquitous part of even remote Ladakhi landscapes. *Source*: Photograph by Lanka Nanayakkara.

ones every New Year. Ladakhi New Year is called Losar, and occurs in December, involving great celebrations, festivities, and family time. Mani walls are another common presence within the landscape. They are around 1 m high and 1–2 m wide, built of cemented stone, filled with mani stones, and decorated with prayer flags. Large prayer wheels may be located at the ends of the wall. Mani walls are linear and can extend for several kilometres. Many villages have mani walls, within the village or at the entrance/exit to the village. Padum and Leh both have long mani walls. Pilgrims walk and circumnavigate mani walls. Prayer wheels come in all sizes. Some are carried by hand, and others are large 2–4 m high structures, set within a plinth, around which pilgrims walk whilst turning the wheel. Inside the wheel are many sacred mantras and texts. They are consecrated and become parts of the village, or town, or even within quite remote locations. At different times of the day, people walk around the prayer wheel. Drivers sometimes stop at special/auspicious prayer wheels and make one or three rounds, then return to their car. Within houses, many people hold prayer wheels and turn them ceaselessly for hours on end.

What becomes astoundingly clear to people who spend time in Ladakh, particularly if they walk around towns, villages, landscapes, and gompas/monasteries, is the realisation that every part of the landscape is touched by the sacred. If you walk on a busy street, you will come across a mani wall or numerous prayer wheels. If you walk in a remote valley, you will see whitened chortens of varying states, new, old, and dilapidated. If you climb high paths and passes, you will encounter hundreds of prayer flags marking the pass or the la, or another prominent feature. You may stare at a cliff, and, after some time of staring, notice a statue carved into the cliff face. The whole tectonic landscape becomes some new reality, with the many marks of Tibetan Buddhism celebrating the spirit of Ladakh.

EPILOGUE

Authors' Travel Vignettes

The authors were privileged to witness much of Ladakh's culture and beauty. We believe an authentic way to communicate Ladakhi culture as we found it is by sharing some of our impressions for readers through short accounts of various locations in Ladakh, narrated by Lanka Nanayakkara. Figure 8.20 presents a story in itself of a Tibetan refugee in Leh, selling a wide range of artefacts from Tibet to support himself and his family: during the 1960s, this man was one of many who walked across the snowy Tibetan mountains to refuge in Ladakh. In the accounts below, Michael refers to author Michael Petterson.

SECMOL: A School in Ladakh

Inaugurated in 1998 by the Dalai Lama, a brainchild of the innovative and famous Ladakhi, Sonam Wangchuk, SECMOL, located 10 km west of Leh, is a school for students, mostly aged 16–19 years. Many students are from disadvantaged families who have struggled to achieve within the normal education system but now find their own space in which to develop. SECMOL (Students' Educational and Cultural Movement of Ladakhis) is funded through charitable subscriptions and does not charge for its residential students. The institution is built using simple, low-cost, traditional building materials and techniques, and follows a 100% self-sustainable model. SECMOL policy on eco-friendly living (especially energy efficiency and handling of waste) is most impressive, presenting caring for the environment in action. At SECMOL, students take 90% of the responsibility for running the school. Their curriculum is a mix of academic and life skills. Their motto is 'learning by doing'. We were lucky to volunteer here for a few weeks. Students come from rural villages and less-advantaged backgrounds, from every corner of Ladakh. Getting to know them and spending time with them was a real privilege for us. We enjoyed

FIGURE 8.20 Tibetan refugee selling a wide range of Tibetan arts and jewellery within a Tibetan market in Leh Source: Photograph by Lanka Nanayakkara.

every minute spent with these kind, happy, bubbly, respectful, and responsible young people. From 6.30 am to 8.30 pm, students have a busy, structured timetable. Apart from studying for their formal curriculum, involving science, maths, environment, engineering, cultural and social studies, English language, etc., they engage in individual-based projects and conversational classes. Each student is assigned daily responsibilities such as looking after/milking the school cows, ensuring that solar batteries are charged and power is not overused, helping out in the kitchen, segregating and sorting waste streams, scrubbing bathrooms, growing vegetables and maintaining the garden, cleaning toilets, dormitories, and halls, to name but a few activities. In mid-winter, they even build their own ice hockey rink! There is time for lots of fun, playing sports, enjoying company, engaging in debates, singing songs, and playing music. A wonderful opportunity for these rural-based students.

Turtuk (Near -CeaseFire Line with Pakistan, in Nubra)

Staring is not considered rude in Asia: it is, in fact, a pastime. Wherever we went, starers stared with enthusiasm. In a country where people hardly marry outside their religion, caste, or ethnicity, we (as a mixed ethnic couple) are a mind-boggling sight to onlookers. On top of that, Michael's silver locks attracted attention in airports, street markets, alleyways, bazaars, and gompas. The majority of Indian men blacken their silver locks with black shoe polish, like dye. Many Indians thought Michael looked like the Missile Man of India: the late Dr Abdul Kalam, the small-town boy who ended up being a rocket scientist, then chief science advisor to the Indian government, and finally the 11th Indian President. He was one of those Indians loved and admired by young and old, southerners and northerners, Hindus and Muslims alike. While wandering down a narrow lane in a dot-like hamlet in the mountains, we were 'ambushed' by a group of local lads. They were present in force. A wall. A barricade. And they started launching endless questions at us. 'Where you from?' Always the first question. 'New Zealand'. I said, 'Do you know where New Zealand is?' I batted back. All shook their heads in unison. 'Is it in Australia?' a smart one jumped in. I was glad my Kiwi friends in Kiwi land were not around to hear such sacrilegious words. So, I took my phone out, showed them the two small islands halfway around the map, almost at the bottom. The little eyes and open mouths inspected it for two and a half seconds in silence. I tried to explain about the ocean and sea, but how do you explain the sea to a kid (or a grown-up) who had only lived at an elevation of more than 3000 metres above sea level and never seen the sea? For them, the nearest beach is over 2,000 km away. So, I showed them photos of our house by the water and the Waitemata Harbour, Auckland. The gang jostled around me now, even more intrigued. 'Ah!...sea' a sudden exclamation. 'Lots of sharks in the sea', 'Shark attack you?' asked the cute one with curious eyes. Others supported his theory – 'there's a lot of shark in the sea'. Meanwhile, the tall cheeky one, now obviously tired of geography talks, wanted to touch Michael's hair and pull it a bit to check if it's real. In quick succession, a few others joined him. White man's long snow-like hair seems much more intriguing than a sea full of sharks thousands of miles away. And once again, we were re-educated about 'Abdul Kalam', 'rocket scientist'.

Processing Barley and Delivering Mail, Ladakh Style

We stayed in a village called Pibiting, part of Zanskar, with a welcoming home-stay family, and were invited to locals processing barley, post-harvest. A delighted meme ley and abi ley (grandpa and grandma) showed us their bountiful harvest from their own fields. Barley is the key crop in Tibet and neighbouring Ladakh. Tibetan people have grown barley for thousands of years. The crop grows well in harsh high altitudes of the Tibetan Plateau. Tsampa and Namkeen Chai (roasted barley and butter tea) are staple Ladakhi foods. Meme ley is a retired mail runner who used to carry a sack full of letters and government documents. He walked on foot, delivering mail to remote villages across Zanskar and towards Leh. During the summer, he would traverse harsh desert-like terrains, with the sun beating down on him. In winter, he would walk like a penguin on frozen rivers in Arctic-like conditions. Some trips only took 8 hours; others, eight days. He would sleep outside with his tsampa and butter tea.

Homestay Experience, Pibiting, Padum, Zanskar

Our first accommodation in Zanskar was a beautiful homestay run by a traditional Ladakhi family. Staying with a family of eight, parents, four young sons, meme ley and abi ley (grandparents), gave us a fascinating insight into the Ladakhi lifestyle and cultural nuances. We also had fresh milk in our tea: they have six cows! Sherab (landlady and mother) milks the cows twice daily. One day, I was sitting in their kitchen-cum-living-room, watching Sherab preparing local food for dinner. Tonyot (one of the twins) had returned from school and sulked in a corner, feeling sorry for himself, his eyes full of large tear balls. With four young boys, all under ten, there was heaps of sibling rivalry going on. Just to divert his thoughts or help him snap out of his woeful mood, I said to him, 'Tonyot, would you like to play chess?'. His face changed at once. He scanned me with his large eyes, half hopeful, half confused. The little guy had never heard of chess. So that night, after dinner, the three elder brothers, Gyaltsen, Tonyot, and Namdak, had their first introduction to the game of chess. The three brothers were absolutely fascinated by our little magnetic travel chess set. All went well until the youngest one arrived like a tornado. He realised he had been left out of whatever the elder brothers were up to and decided that it was time to wreck the chess game. Chess pieces flew everywhere! Luckily, at other times, the family enjoyed many (more peaceful) chess games, playing on top of a chokse (a low Ladakhi traditional ornate table).

A Nunnery in Zangla, Zanskar

Why do nunneries get significantly less attention and funding compared with monasteries? Why are women's contributions undervalued in religious institutions as in many fields of science, politics, performing arts, and life in general? In support of sisterhood, while we were in Zanskar, we popped

FIGURE 8.21 A group of nuns cleaning their brassware used for Buddhist rituals within Dorjey Zong Nunnery, founded in the 14th century CE. Source: Photograph by Lanka Nanayakkara.

into a few nunneries whenever time permitted. Zanskar is home to around ten nunneries, some quite tricky to reach. They are usually situated in remote places, up terrain, along winding gravel roads, with no signboards to announce their presence. Nunneries don't receive many local visitors. Foreign tourists are rare. The nuns were very sweet to us. We always received warm smiles and welcomes, with unlimited steaming cups of tea. Although language was an annoying barrier, we tried to engage with them, and our guide and driver, Yangjor, came in handy. At the tender age of 89, Tsering Chuskit is the oldest nun in the Zangla Buddhist nunnery. The first time we met her, I asked her how she finds life in the nunnery. All nuns use the local (squatting earth) toilet, some 20–30 m away from their living quarters. So, for Tsering Chuskit, the trip to the loo involves a fair bit of walking through a neat flower bed, down steps, then across an open courtyard, around outer buildings, more steps, and finally the toilet! Phew! Funnily, she didn't complain about the outside toilet; instead mentioned that her ears hurt in winter. Winters in Zanskar are harsh. Between December and March, temperatures drop to below zero, sometimes reaching minus 20 or 30°C. We managed to get Tsering some warm clothes for the winter. Figure 8.21 presents a group of nuns cleaning brassware, one of many tasks to keep a nunnery working.

Sand Mandalas at Chemrey Gompa, Indus Valley

When we visited Chemrey gompa, the lamas (Tibetan monks) were preparing a sand mandala for an upcoming puja (Figure 8.16). We have seen many 3D mandalas, wall paintings, and thangka paintings of mandalas, but never before witnessed a 2D sand mandala. Mandalas are symbolic representations of the universe, Tibetan Buddhist style, depicting different teachings of Buddhism. The process begins with marking the design with chalk on a flat surface. Then, starting from the centre, monks painstakingly work their way outwards, filling in the design with coloured sand (crushed natural minerals). We've been told it usually takes around 6–7 days to complete a complex mandala. When completed, the mandala is consecrated, then destroyed. The monks carefully sweep coloured sand from the perimeter to the centre, collect all the sand grains in an urn, and finally disperse the particles in flowing water. Destroying it, leaving no trace, symbolises the transitory nature of all things in the universe. Buddhist values of non-attachment and impermanence are manifested.

Lingshed, High Mountain Valley in Zanskar

After travelling for 5–6 hours along bumpy, nerve-racking roads, we finally reached Lingshed at around 4 pm. We had never anticipated finding a village tucked away amidst such desolate mountain peaks, so high, so isolated, even by Ladakhi standards. According to Google Maps, there are only four homestays in the whole of Lingshed, but we were unable to call or pre-book a place. Once we reached the village, we paid our respects at the old Gompa, and we went around the village searching for these four accommodations. After a few inquiries, we managed to find two places, but they were already full, and the 3rd and 4th ones were nowhere to be found. When all options seemed to be slipping away in the dark, anxiety levels and hunger pangs creeping inside, a shepherd heading home with his flock stopped for a minute to inform us of yet another place down a small valley (which looked more like a dark precipice and a steep drop). Fingers and toes crossed, we followed a steep dusty gravel road, and, all of a sudden, a board announcing a 'HOMASTEY' came into sight! Then a small, modest house veiled in dark, with a low door, a faint light, and a hunched, toothless lady, flashing a warm, prodigious smile, appeared out of the darkness. It felt like a scene from the Russian folktale book, *Vasilisa the Beautiful*. That's how we ended up in Dorjey Namgail Homestay. The old lady smelled of history. During our 15 hours under her roof, I managed to charm her and extract a bit of her life story. Going back 60 years, Tsering Palmo (Figure 8.22) grew up in the prosperous Zangla village in Zanskar. Zangla boasts a fertile valley, an ancient palace, an influential monarchy, and a proud heritage. One day, a man from Lingshed appeared with gifts such as tsampa, yak butter, and copper pots. He asked her parents for Tsering Palmo's hand in marriage. Soon it was decided by the elders that she would be married off to Lingshed. The wedding party, including yaks, goats, donkeys, and horses (wedding gifts from her parents), ambled over mountain passes and valleys for 3–4 days, finally reaching her new home in Lingshed. She can't recall the exact year nor her age. She did remember that Lingshed only had one hundred houses when she arrived. Her husband owned seven yaks, four horses, many cows, sheep, goats, and donkeys. He was part of a local trading caravan and was away regularly for long periods of time. Soon, Tsering Palmo was afflicted with homesickness. She 'cried and cried and cried' and ran back to Zangla. But as Arundhati Roy's Booker Prize-*winning The God of Small Things* talks about 'the commonly held view that a married daughter had no position in her parents' home', Tsering Palmo's parents escorted her back to her husband's village. There was no question about where she belonged. Fast track many moons, here she is, aged 77 years old, a widow for 20 years, with five children, four of whom are married off, and one son, a Geshe La (PhD level) monk in Karnataka. She tells us, very proudly, that she has been to several Buddhist pilgrimage sites, including Lumbini in Nepal, and is looking forward to attending her Geshe-son's graduation in Karnataka at the end of this year.

Phuktal, Zanskar

Phuktal monastery is one of the few gompas in Ladakh that can be reached only on foot (Figure 8.15). There are two treks: an easy flat one following the river valley, and a narrow winding trail along steep mountain tops with vertical slopes, allowing for breathtaking views of the deep gorge and the turquoise river below. Guess which one we took? Our trek started from Purne village (3,950 m above sea level),

FIGURE 8.22 Tsering Palmo from Lingshed dressed in a traditional Ladakhi cold weather outfit, holding her mala beads and reciting mantras. Source: Photograph by Lanka Nanayakkara.

with a 2-hour trek in the early morning, mostly downhill, being a breeze. Coming back was a different story. We stopped a fair number of times, gasping for air, lungs working overtime, as if oxygen was a rare commodity (which it is for us in Ladakh). Nestled in the remote Lungnak Valley, Phuktal is an architectural marvel, situated within a stunning mountain setting. The gompa is built high on the inside of a cliff that overhangs a deep gorge. The whole complex resembles a honeycomb. Established during the early 12th century CE, the natural cave inside has been used by arahants, gurus, sages, yogis, and monks for a millennium.

FIGURE 8.23 Looking to the future. Two young women dressed in traditional outfits, resplendent with jewellery and adornments, prepared for performances within an outdoor art festival, with the latest modern technology in hand. Source: Photgraph by Lanka Nanayakkara.

REFERENCES

Bedi, R. S. (1986). *Ladakh*. New Delhi: Brijbasi Publications Pvt Ltd., 112pp.

Chauhan, A., Jishtu, V., Thakur, L., & Dolma, T. (2020). Medicinal plants of the Trans-Himalayan cold desert of Ladakh – A review. *International Journal of Science, Environment and Technology*, 9(2), 239–253.

Chernicoff, S., & Whitney, D. (2007). *Geology: An Introduction to Physical Geology* (4th ed.). Pearson Prentice Hall, p. 585. ISBN 9780131751248.

Chumikchan, R. A. (2019). Place with a heart for animals. Reach Ladakh Bulletin. https://www.reachladakh.com/news/oped-page/a-place-with-a-heart-for-animals.

Deshpande, A. (2024). *Speaking Stones. Rock Art of Ladakh.* Published by A. Deshpande. Hyderabad: Printed by Pragati Offset Pvt., Ltd.

Directorate of Tourism, Union Territory of Ladakh. (2024). *Year-Wise Tourist Arrivals in Ladakh (2005–2023).* https://leh.nic.in/tourism/.

District Leh. (2026, February 27). *History.* Union Territory of Ladakh. https://leh.nic.in/about-district/history/.

Douglas, E. (2021). *Himalaya: A Human History.* London, UK: Penguin Books.

Douglas, E. (2003). *Tenzing, Hero of Everest.* Washington, DC: National Geographic, 299 pages.

Fatland, E. (2022). *High. A Journey Across the Himalayas Through Pakistan, India, Bhutan, Nepal and China.* London, UK: Quercus Publications.

Gulia, S., & Singh, G. (2021). High-altitude adaptation: Role of genetic and epigenetic factors. *Molecular Biology Reports*, 48, 7531–7543.

Gupta, R. D., & Sanjay, A. (2017). Characteristics of the soils of Ladakh region of Jammu and Kashmir. *Journal of Soil and Water Conservation,* 16(3), 260–266. ISSN: 022-457X (Print); 2455-7145 (Online); https://doi.org/10.5958/2455-7145.2017.00037.6.

Harvey, A. (1983/1984/1993). *A Journey in Ladakh.* London: Jonathan Cape/HarperCollins/Picador Publications.

Harvey, P. (1990 and 2004) (different updated editions). *An Introduction to Buddhism. Teachings, History, and Practices.* Cambridge, UK: Cambridge University Press, 374 pp.

Kaul, H. N. (1992). *Ladakh Through the Ages, Towards a New Identity.* New Delhi: Indus Publishing Group.

Kaul, H. N. (1998). *Rediscovery of Ladakh.* New Delhi: Indus Publishing Group.

Keay, J. (1979). *The Gilgit Game. The Explorers of the Western Himalayas 1865–95.* Oxford, UK: Oxford University Press, 290pp.

Krinsley, D., Dorn, R., & Tovey, N. K. (1995). Nanometer-scale layering in rock varnish: Implications for genesis and paleoenvironmental interpretation. *The Journal of Geology,* 103(1), 106–113. https://doi.org/10.1086/629726. S2CID 129392419.

Ladakhdekho.com. (2026). Traditional medicine in Ladakh. https://www.ladakhdekho.com/traditional-medicine-in-ladakh.html (accessed March 2026).

Marley, B., & the Wailers. (1983). Confrontations. Island Records. ILPS 9760.

Men-Tsee-Khang Sowa-Rigpa Medical College & Hospital, Dharamsala. https://dcollege.mentseekhang.org/.

National Institute of Sowa Rigpa. https://www.sowarigpainstitute.in/.

Neetu, D. J. S. (1994). *Ladakh.* New Delhi: Brijbasi Printers Pvt Ltd., 152pp.

Norberg-Hodge, H. (1991). *Ancient Futures. Learning from Ladakh.* Oxford, UK: Oxford University Press, 204pp.

Office of the Registrar General & Census Commissioner, India. (2013). Census of India 2011: District census handbook, Leh (Ladakh). Ministry of Home Affairs. censusindia.gov.in.

Perry, R. S., & Adams, J. B. (1978). Desert varnish: Evidence for cyclic deposition of manganese. *Nature,* 276(5687), 489–491. https://doi.org/10.1038/276489a0. S2CID 4318328.

Shakspo, N. T. (2024). *A Cultural History of Ladakh.* Leh-Ladakh: Centre for Research on Ladakh, 299pp.

Singh, J. (2025). Status of the natural environment in Ladakh. Centre for Excellence for Himalayan Studies. RGICS. https://www.rgics.org/environment/status-of-the-natural-environment-in-ladakh/.

Index

For Product Safety Concerns and Information please contact our EU representative GPSR@taylorandfrancis.com Taylor & Francis Verlag GmbH, Kaufingerstraße 24, 80331 München, Germany

Batch number: 10784218

Printed by Printforce, the Netherlands